Mechanical Engineering Series

Frederick F. Ling
Series Editor

Springer
*New York
Berlin
Heidelberg
Barcelona
Hong Kong
London
Milan
Paris
Singapore
Tokyo*

Mechanical Engineering Series

J. Angeles, **Fundamentals of Robotic Mechanical Systems:
Theory, Methods, and Algorithms**

P. Basu, C. Kefa, and L. Jestin, **Boilers and Burners: Design and Theory**

J.M. Berthelot, **Composite Materials:
Mechanical Behavior and Structural Analysis**

I.J. Busch-Vishniac, **Electromechanical Sensors and Actuators**

J. Chakrabarty, **Applied Plasticity**

G. Chryssolouris, **Laser Machining: Theory and Practice**

V.N. Constantinescu, **Laminar Viscous Flow**

G.A. Costello, **Theory of Wire Rope, 2nd ed.**

K. Czolczynski, **Rotordynamics of Gas-Lubricated Journal Bearing Systems**

M.S. Darlow, **Balancing of High-Speed Machinery**

J.F. Doyle, **Wave Propagation in Structures:
Spectral Analysis Using Fast Discrete Fourier Transforms, 2nd ed.**

P.A. Engel, **Structural Analysis of Printed Circuit Board Systems**

A.C. Fischer-Cripps, **Introduction to Contact Mechanics**

J. García de Jalón and E. Bayo, **Kinematic and Dynamic Simulation of
Multibody Systems: The Real-Time Challenge**

W.K. Gawronski, **Dynamics and Control of Structures: A Modal Approach**

K.C. Gupta, **Mechanics and Control of Robots**

J. Ida and J.P.A. Bastos, **Electromagnetics and Calculations of Fields**

M. Kaviany, **Principles of Convective Heat Transfer**

M. Kaviany, **Principles of Heat Transfer in Porous Media, 2nd ed.**

E.N. Kuznetsov, **Underconstrained Structural Systems**

P. Ladevèze, **Nonlinear Computational Structural Mechanics:
New Approaches and Non-Incremental Methods of Calculation**

A. Lawrence, **Modern Inertial Technology: Navigation, Guidance, and
Control, 2nd ed.**

R.A. Layton, **Principles of Analytical System Dynamics**

C.V. Madhusudana, **Thermal Contact Conductance**

(continued after index)

Anthony C. Fischer-Cripps

Introduction to Contact Mechanics

With 93 Figures

 Springer

Anthony C. Fischer-Cripps
CSIRO
Bradfield Road, West Lindfield
Lindfield NSW 2070, Australia

Series Editor
Frederick F. Ling
Ernest F. Gloyna Regents Chair
  in Engineering
Department of Mechanical Engineering
The University of Texas at Austin
Austin, TX 78712-1063, USA
  and
William Howard Hart Professor Emeritus
Department of Mechanical Engineering,
  Aeronautical Engineering and Mechanics
Rensselaer Polytechnic Institute
Troy, NY 12180-3590, USA

Consulting Editor
Iain Finnie
Department of Mechanical Engineering
University of California
Berkeley, CA 94720, USA

Library of Congress Cataloging-in-Publication Data
Fischer-Cripps, Anthony C.
    Introduction to contact mechanics / Anthony C. Fischer-Cripps.
        p.   cm. — (Mechanical engineering series)
    Includes bibliographical references and index.
    ISBN 0-387-98914-5 (hc. : alk. paper)
    1. Contact mechanics. I. Title. II. Series: Mechanical
  engineering series (Berlin, Germany)
  TA353.F57   2000
  620.1—dc21   99-42671

Printed on acid-free paper.

© 2000 Springer-Verlag New York, Inc.
All rights reserved. This work may not be translated or copied in whole or in part without the written permission of the publisher (Springer-Verlag New York, Inc., 175 Fifth Avenue, New York, NY 10010, USA), except for brief excerpts in connection with reviews or scholarly analysis. Use in connection with any form of information storage and retrieval, electronic adaptation, computer software, or by similar or dissimilar methodology now known or hereafter developed is forbidden.
The use of general descriptive names, trade names, trademarks, etc., in this publication, even if the former are not especially identified, is not to be taken as a sign that such names, as understood by the Trade Marks and Merchandise Marks Act, may accordingly be used freely by anyone.

Production coordinated by Robert Wexler and managed by Francine McNeill; manufacturing supervised by Jeffrey Taub.
Photocomposed copy prepared from the author's Microsoft Word files.
Printed and bound by Edwards Brothers, Inc., Ann Arbor, MI.
Printed in the United States of America.

9 8 7 6 5 4 3 2 1

ISBN 0-387-98914-5   Springer-Verlag New York Berlin Heidelberg   SPIN 10742874

To Dianne and Raymond

Mechanical Engineering Series

Frederick F. Ling
Series Editor

Advisory Board

Applied Mechanics	F.A. Leckie University of California, Santa Barbara
Biomechanics	V.C. Mow Columbia University
Computational Mechanics	H.T. Yang University of California, Santa Barbara
Dynamical Systems and Control	K.M. Marshek University of Texas, Austin
Energetics	J.R. Welty University of Oregon, Eugene
Mechanics of Materials	I. Finnie University of California, Berkeley
Processing	K.K. Wang Cornell University
Production Systems	G.-A. Klutke Texas A&M University
Thermal Science	A.E. Bergles Rensselaer Polytechnic Institute
Tribology	W.O. Winer Georgia Institute of Technology

Series Preface

Mechanical engineering, an engineering discipline forged and shaped by the needs of the industrial revolution, is once again asked to do its substantial share in the call for industrial renewal. The general call is urgent as we face profound issues of productivity and competitiveness that require engineering solutions. The Mechanical Engineering Series features graduate texts and research monographs intended to address the need for information in contemporary areas of mechanical engineering.

The series is conceived as a comprehensive one that covers a broad range of concentrations important to mechanical engineering graduate education and research. We are fortunate to have a distinguished roster of consulting editors on the advisory board, each an expert in one of the areas of concentration. The names of the consulting editors are listed on the facing page of this volume. The areas of concentration are applied mechanics, biomechanics, computational mechanics, dynamic systems and control, energetics, mechanics of materials, processing, production systems, thermal science, and tribology.

Professor Finnie, the consulting editor for mechanics of materials, and I are pleased to present *Introduction to Contact Mechanics* by Anthony C. Fischer-Cripps.

Austin, Texas   Frederick F. Ling

Preface

This book deals with the mechanics of solid bodies in contact, a subject intimately connected with such topics as fracture, hardness, and elasticity. Theoretical work is most commonly supported by the results of indentation experiments under controlled conditions. In recent years, the indentation test has become a popular method of determining mechanical properties of both brittle and ductile materials, and particularly thin film systems.

The book begins with an introduction to the mechanical properties of materials, general fracture mechanics, and the fracture of brittle solids. This is followed by a detailed description of indentation stress fields for both elastic and elastic-plastic contact. The discussion then turns to the formation of Hertzian cone cracks in brittle materials, subsurface damage in ductile materials, and the meaning of hardness. The book concludes with an overview of practical methods of indentation testing.

My intention is for this book to make contact mechanics accessible to those entering the field for the first time. Experienced researchers may also benefit from the review of the most commonly used formulas and theoretical treatments of the past century.

In writing this book, I have been assisted and encouraged by many colleagues, friends, and family. I am most indebted to A. Bendeli, R.W. Cheary, R.E. Collins, R. Dukino, J.S. Field, A.K. Jämting, B.R. Lawn, C.A. Rubin, and M.V. Swain. Finally, I thank Dr. Thomas von Foerster and the production team at Springer-Verlag New York, Inc., for their very professional and helpful approach to the whole publication process.

Lindfield, Australia   Anthony C. Fischer-Cripps

Contents

Series Preface	vii
Preface	ix
List of Symbols	xvii
History	xix

Chapter 1. Mechanical Properties of Materials ... 1

  1.1 Introduction ... 1
  1.2 Elasticity ... 1
    1.2.1 Forces between atoms ... 1
    1.2.2 Hooke's law ... 2
    1.2.3 Strain energy ... 4
    1.2.4 Surface energy ... 4
    1.2.5 Stress ... 5
    1.2.6 Strain ... 10
    1.2.7 Poisson's ratio ... 14
    1.2.8 Linear elasticity (generalized Hooke's law) ... 14
    1.2.9 2-D Plane stress, plane strain ... 16
    1.2.10 Principal stresses ... 18
    1.2.11 Equations of equilibrium and compatibility ... 23
    1.2.12 Saint-Venant's principle ... 25
    1.2.13 Hydrostatic stress and stress deviation ... 25
    1.2.14 Visualizing stresses ... 26
  1.3 Plasticity ... 27
    1.3.1 Equations of plastic flow ... 27
  1.4 Stress Failure Criteria ... 28
    1.4.1 Tresca failure criterion ... 28
    1.4.2 Von Mises failure criterion ... 29
  References ... 30

Chapter 2. Linear Elastic Fracture Mechanics ... 31

  2.1 Introduction ... 31
  2.2 Stress Concentrations ... 31

Contents

2.3 Energy Balance Criterion 32
2.4 Linear Elastic Fracture Mechanics 37
    2.4.1 Stress intensity factor 37
    2.4.2 Crack tip plastic zone 40
    2.4.3 Crack resistance 41
    2.4.4 K_{1C}, the critical value of K_1 41
    2.4.5 Equivalence of G and K 42
2.5 Determining Stress Intensity Factors 43
    2.5.1 Measuring stress intensity factors experimentally 43
    2.5.2 Calculating stress intensity factors from prior stresses 44
    2.5.3 Determining stress intensity factors
       using the finite-element method 46

References 48

Chapter 3. Delayed Fracture in Brittle Solids 49

3.1 Introduction 49
3.2 Static Fatigue 49
3.3 The Stress Corrosion Theory of Charles and Hillig 51
3.4 Sharp Tip Crack Growth Model 54
3.5 Using the Sharp Tip Crack Growth Model 57

References 59

Chapter 4. Statistics of Brittle Fracture 61

4.1 Introduction 61
4.2 Basic Statistics 62
4.3 Weibull Statistics 64
    4.3.1 Strength and failure probability 64
    4.3.2 The Weibull parameters 66
4.4 The Strength of Brittle Solids 68
    4.4.1 Weibull probability function 68
    4.4.2 Determining the Weibull parameters 70
    4.4.3 Effect of biaxial stresses 71
    4.4.4 Determining the probability of delayed failure 74

References 76

Chapter 5. Elastic Indentation Stress Fields 77

5.1 Introduction 77
5.2 Hertz Contact Pressure Distribution 77
5.3 Analysis of Indentation Stress Fields 78
    5.3.1 Line contact 79

  5.3.2 Point contact 80
  5.3.3 Analysis of stress and deformation 82
 5.4 Indentation Stress Fields 83
  5.4.1 Uniform pressure 84
  5.4.2 Spherical indenter 87
  5.4.3 Cylindrical roller (2-D) contact 92
  5.4.4 Cylindrical (flat punch) indenter 93
  5.4.5 Rigid cone 97
 References 101

Chapter 6. Elastic Contact 103
 6.1 Hertz Contact Equations 103
 6.2 Contact Between Elastic Solids 104
  6.2.1 Spherical indenter 105
  6.2.2 Flat punch indenter 109
  6.2.3 Conical indenter 109
 6.3 Impact 110
 6.4 Friction 112
 References 116

Chapter 7. Hertzian Fracture 117
 7.1 Introduction 117
 7.2 Hertzian Contact Equations 117
 7.3 Auerbach's Law 118
 7.4 Auerbach's Law and the Griffith Energy Balance Criterion 119
 7.5 Flaw Statistical Explanation of Auerbach's Law 120
 7.6 Energy Balance Explanation of Auerbach's Law 120
 7.7 The Probability of Hertzian Fracture 126
  7.7.1 Weibull statistics 126
  7.7.2 Application to indentation stress field 127
 7.8 Fracture Surface Energy and the Auerbach Constant 131
  7.8.1 Minimum critical load 131
  7.8.2 Median fracture load 133
 7.9 Cone Cracks 135
  7.9.1 Crack path 135
  7.9.2 Crack size 136
 References 136

Chapter 8. Elastic-Plastic Indentation Stress Fields 139
 8.1 Introduction 139

8.2 Pointed Indenters 139
   8.2.1 Indentation stress field 139
   8.2.2 Indentation fracture 144
   8.2.3 Fracture toughness 146
   8.2.4 Berkovich indenter 147
8.3 Spherical Indenter 148
References 151

Chapter 9. Hardness 153

9.1 Introduction 153
9.2 Indentation Hardness Measurements 153
   9.2.1 Brinell hardness number 153
   9.2.2 Meyer hardness 154
   9.2.3 Vickers diamond hardness 155
   9.2.4 Knoop hardness 155
   9.2.5 Other hardness test methods 156
9.3 Meaning of Hardness 156
   9.3.1 Compressive modes of failure 157
   9.3.2 The constraint factor 158
   9.3.3 Indentation response of materials 159
   9.3.4 Hardness theories 160
References 174

Chapter 10. Elastic and Elastic-Plastic Contact 177

10.1 Introduction 177
10.2 Geometrical Similarity 177
10.3 Indenter Types 178
   10.3.1 Spherical, conical, and pyramidal indenters 178
   10.3.2 Sharp and blunt indenters 181
10.4 Elastic-Plastic Contact 181
   10.4.1 Elastic recovery 181
   10.4.2 Compliance 185
   10.4.3 Analysis of compliance curves 186
   10.4.4 The elastic-plastic contact surface 196
10.5 Internal Friction and Plasticity 197
References 199

Chapter 11. Indentation Test Methods 201

11.1 Introduction 201
11.2 Bonded-Interface Technique 201

11.3 Indentation Stress-Strain Response 203
    11.3.1 Theoretical 203
    11.3.2 Experimental method.. 204
11.4 Compliance Curves 207
11.5 Inert Strength... 209
11.6 Hardness Testing 212
    11.6.1 Vickers hardness.. 212
    11.6.2 Berkovich indenter 214
    11.6.3 Knoop hardness... 214
References 215

Appendix 1. Submicron Indentation Test Analysis 217

A1.1 Introduction 217
A1.2 Initial Penetration Depth... 217
A1.3 Instrument Compliance... 219
A1.4 Indenter Shape Correction 221
A1.5 Hardness as a Function of Depth 224
A1.6 Generating Simulated Data... 226
    A1.6.1 Berkovich indenter 226
    A1.6.2 Spherical indenter.. 227
References 230

Appendix 2. The Finite-Element Method 231

A2.1 Introduction 231
A2.2 Finite-Element Analysis 231
A2.3 Finite-Element Modeling.. 234
    A2.3.1 Contact between the indenter and the specimen 234
    A2.3.2 Elastic-plastic response 235
    A2.3.3 Finite-element model... 237
    A2.3.4 Finite-element modeling results... 238
References 239

Index 241

List of Symbols

a	cylindrical indenter radius or spherical indenter contact area radius
α	cone semi-angle
A	Auerbach constant; area; material characterization factor
b	distance along a crack path
B	risk function
β	friction parameter; rate of stress increase; cone inclination angle, indenter shape factor
c	total crack length; radius of elastic-plastic boundary
c_o	size of plastic zone
C	hardness constraint factor, compliance
δ	distance of mutual approach between indenter and specimen
d	length of long diagonal
D	subcritical crack growth constant; spherical indenter diameter
E	Young's modulus
E_o	activation energy
ε	strain
F	force
G	strain energy release rate per unit of crack extension; shear modulus
h	plate thickness; distance; indentation depth
H	hardness
I	matrix subscript
j	matrix subscript
κ	stress concentration factor
k	Weibull strength parameter; elastic spring stiffness constant; Boltzmann's constant, elastic mismatch parameter, initial depth constant
K	bulk modulus, Oliver and Pharr correction factor
K_1	stress intensity factor for mode 1 loading.
K_{1scc}	static fatigue limit
L, l	length or distance
λ	Lamé constant
m	Weibull modulus
n	subcritical crack growth exponent; number; ratio of minimum to maximum stress, initial depth exponent

List of Symbols

N	total number
η	coefficient of viscosity
P	indenter load (force)
P_f	probability of failure
p_m	mean contact pressure
P_s	probability of survival
ϕ	strain energy release function
q	uniform lateral pressure
θ	angle
R	universal gas constant; spherical indenter radius
r	radial distance
RH	relative humidity
r_o	ring crack starting radius
ρ	radius of curvature; number density
s	distance
σ	normal stress
T	temperature
t	time
τ	shear stress
u	displacement
U	energy
μ	Lamé constant, coefficient of friction
V	volume
ν	Poisson's ratio
W	work
x	linear displacement, strain index
γ	surface energy; shear angle
γ_o	activation energy
Y	yield stress, shape factor

History

It may surprise those who venture into the field of "contact mechanics" that the first paper on the subject was written by Heinrich Hertz. At first glance, the nature of the contact between two elastic bodies has nothing whatsoever to do with electricity, but Hertz recognized that the mathematics was the same and so founded the field, which has retained a small but loyal following during the past one hundred years.

Hertz wanted to be an engineer. In 1877, at age 20, he traveled to Munich to further his studies in engineering, but when he got there, doubts began to occupy his thoughts. Although "there are a great many sound practical reasons in favor of becoming an engineer" he wrote to his parents, "I still feel that this would involve a sense of failure and disloyalty to myself." While studying engineering at home in Hamburg, Hertz had become interested in natural science and was wondering whether engineering, with "surveying, building construction, builder's materials and the like," was really his lifelong ambition. Hertz was really more interested in mathematics, mechanics, and physics. Guided by his parents' advice, he chose the physics course and found himself in Berlin a year later to study under Hermann von Helmholtz and Gustav Kirchhoff.

In October 1878, Hertz began attending Kirchhoff's lectures and observed on the notice board an advertisement for a prize for solving a problem involving electricity. Hertz asked Helmholtz for permission to research the matter and was assigned a room in which to carry out experiments. Hertz wrote: "every morning I hear an interesting lecture, and then go to the laboratory, where I remain, barring a short interval, until four o'clock. After that, I work in the library or in my rooms." Hertz wrote his first paper, "Experiments to determine an upper limit to the kinetic energy of an electric current," and won the prize.

Next, Hertz worked on "The distribution of electricity over the surface of moving conductors," which would become his doctoral thesis. This work impressed Helmholtz so much that Hertz was awarded "Acuminis et doctrine specimen laudabile" with an added "magna cum laude." In 1880, Hertz became an assistant to Helmholtz—in modern-day language, he would be said to have obtained a three-year "post-doc" position.

On becoming Helmholtz's assistant, Hertz immediately became interested in the phenomenon of Newton's rings—a subject of considerable discussion at the time in Berlin. It occurred to Hertz that, although much was known about the optical phenomena when two lenses were placed in contact, not much was known about the deflection of the lenses at the point of contact. Hertz was particularly concerned with the nature of the localized deformation and the distribution of pressure between the two contacting surfaces. He sought to assign a shape to the surface of contact that satisfied certain boundary conditions worth repeating here:

1. The displacements and stresses must satisfy the differential equations of equilibrium for elastic bodies, and the stresses must vanish at a great distance from the contact surface—that is, the stresses are localized.
2. The bodies are in frictionless contact.
3. At the surface of the bodies, the normal pressure is zero outside and equal and opposite inside the circle of contact.
4. The distance between the surfaces of the two bodies is zero inside and greater than zero outside the circle of contact.
5. The integral of the pressure distribution within the circle of contact with respect to the area of the circle of contact gives the force acting between the two bodies.

Hertz generalized his analysis by attributing a quadratic function to represent the profile of the two opposing surfaces and gave particular attention to the case of contacting spheres. Condition 4 above, taken together with the quadric surfaces of the two bodies, defines the form of the contacting surface. Condition 4 notwithstanding, the two contacting bodies are to be considered elastic, semi-infinite, half-spaces. Subsequent elastic analysis is generally based on an appropriate distribution of normal pressure on a semi-infinite half-space. By analogy with the theory of electric potential, Hertz deduced that an ellipsoidal distribution of pressure would satisfy the boundary conditions of the problem and found that, for the case of a sphere, the required distribution of normal pressure σ_z is:

$$\frac{\sigma_z}{p_m} = -\frac{3}{2}\left(1 - \frac{r^2}{a^2}\right)^{1/2}, \quad r \leq a$$

This distribution of pressure reaches a maximum (1.5 times the mean contact pressure p_m) at the center of contact and falls to zero at the edge of the circle of contact ($r = a$). Hertz did not calculate the magnitudes of the stresses at points throughout the interior but offered a suggestion as to their character by interpolating between those he calculated on the surface and along the axis of symmetry. The full contact stress field appears to have been first calculated in detail by Huber in 1904 and again later by Fuchs in 1913, and by Moreton and Close in 1922. More recently, the integral transform method of Sneddon has been applied to axis-symmetric distributions of normal pressures, which correspond to a variety of indenter geometries. In brittle solids, the most important stress is not the normal pressure but the radial tensile stress on the specimen surface, which reaches a maximum value at the edge of the circle of contact. This is the stress that is responsible for the formation of the conical cracks that are familiar to all who have had a stone impact on the windshield of their car. These cracks are called "Hertzian cone cracks."

Hertz published his work under the title "On the contact of elastic solids," and it gained him immediate notoriety in technical circles. This community interest led Hertz into a further investigation of the meaning of hardness, a field in which he found that "scientific men have as clear, i.e. as vague, a conception as the man in the street." It was appreciated very early on that hardness indicated a resistance to penetration or permanent deformation. Early methods of measuring hardness, such as the scratch method, although convenient and simple, were found to involve too many variables to provide the means for a scientific definition of this property. Hertz postulated that an absolute value for hardness was the least value of pressure beneath a spherical indenter necessary to produce a permanent set at the center of the area of contact. Hardness measurements embodying Hertz's proposal formed the basis of the Brinell test (1900), Shore scleroscope (1904), Rockwell test (1920), Vickers hardness test (1924), and finally the Knoop hardness test (1934).

In addition to being involved in this important practical matter, Hertz also took up researches on evaporation and humidity in the air. After describing his theory and experiments in a long letter to his parents, he concluded with "this has become quite a long lecture and the postage of the letter will ruin me; but what wouldn't a man do to keep his dear parents and brothers and sister from complete desiccation?"

Although Hertz spent an increasing amount of his time on electrical experiments and high voltage discharges, he remained as interested as ever in various side issues, one of which concerned the flotation of ice on water. He observed that a disk floating on water may sink, but if a weight is placed on the disk, it may float. This paradoxical result is explained by the weight causing the disk to bend and form a "boat," the displacement of which supports both the disk and the weight. Hertz published "On the equilibrium of floating elastic plates" and then moved more or less into full-time study of Maxwellian electromagnetics but not without a few side excursions into hydrodynamics.

Hertz's interest and accomplishments in this area, as a young man in his twenties, are a continuing source of inspiration to present-day practitioners. Advances in mathematics and computational technology now allow us to plot full details of indentation stress fields for both elastic and elastic-plastic contact. Despite this technology, the science of hardness is still as vague as ever. Is hardness a material property? Hertz thought so, and many still do. However, many recognize that the hardness one measures often depends on how you measure it, and the area remains as open as ever to scientific investigation.

Chapter 1
Mechanical Properties of Materials

1.1 Introduction

The aim of this book is to provide simple and clear explanations about the nature of contact between solid bodies. It is customary to use the term "indenter" to refer to the body to which the loading force is applied, and to refer to the body undergoing the deformation of interest as the "specimen." Such contact may be purely elastic, or it may involve some plastic, or irreversible, deformation of either the indenter, the specimen, or both. The first two chapters of this book are concerned with the basic principles of elasticity, plasticity, and fracture. It is assumed that the reader is familiar with the engineering meaning of common terms such as force and displacement but not necessarily familiar with engineering terms such as stress, strain, elastic modulus, Poisson's ratio, and other material properties. The aim of these first two chapters is to inform and educate the reader in these basic principles and to prepare the groundwork for subsequent chapters on indentation and contact between solids.

1.2 Elasticity

1.2.1 Forces between atoms

It is reasonable to suppose that the strength of a material depends on the strength of the chemical bonds between its atoms. Generally, atoms in a solid are attracted to each other over long distances (chemical bond forces) and are also repelled by each other at very short distances (Coulomb repulsion). In the absence of any other forces, atoms take up equilibrium positions where these long-range attractions and short-range repulsions balance. The long-range attractive chemical bond forces are a consequence of the lower energy states that arise due to filling of electron shells. The short-range repulsive Coulomb forces are electrostatic in origin.

Figure 1.2.1 shows a representation of the force required to move one atom away from another at the equilibrium position. The exact shape of this relationship depends on the nature of the bond between them (e.g., ionic, covalent, or metallic). However, all bonds show a force–distance relationship of the same general character. As can be seen, *near the equilibrium position*, the force F required to move one atom away from another is very nearly directly proportional to the distance x:

$$F = kx \qquad (1.2.1a)$$

A solid that shows this behavior is said to be "linearly elastic," and this is usually the case for *small displacements* about the equilibrium position for most solid materials. Of course, in reality, the situation is complicated by the effect of neighboring atoms and the three-dimensional character of real solids.

1.2.2 Hooke's law

Referring to Fig. 1.2.1, let us imagine one atom being slowly pulled away from the other by an external force. The maximum value of the external force required to break the chemical bond between them is called the "cohesive strength" To break the bond, at least this amount of force must be applied. From then on, less and less force can be applied until the atom is so far away that very little force is required to keep it there. The strength of the bond, by definition, is equal to the maximum cohesive force.

In general, the shape of the force displacement curve may be approximated by a portion of a sine wave, as shown in Fig 1.2.1. The region of interest is the section from the equilibrium position to the maximum force. In this region,

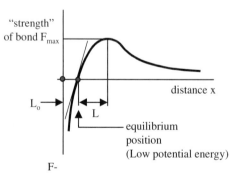
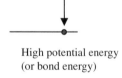

Fig. 1.2.1 Schematic of the forces between atoms in a solid as a function of distance away from the center of the atom. Repulsive force acts over a very short distance. Attractive forces between atoms act over a very long distance. An atom at infinity has a higher potential energy than one at the equilibrium position.

$$F = F_{max} \sin\left(\frac{\pi x}{2L}\right) \qquad (1.2.2a)$$

where L is the distance from the equilibrium position to the position at F_{max}. Now, since $\sin\theta \approx \theta$ for small values of θ, the force required for small displacements x is:

$$\begin{aligned} F &= F_{max} \frac{\pi x}{2L} \\ &= \left[\frac{F_{max}\pi}{2L}\right] x \end{aligned} \qquad (1.2.2b)$$

Now, L and F_{max} may be considered constant for any one particular material. Thus, Eq. 1.2.2b takes the form $F = kx$, which is more familiarly known as Hooke's law. The result can be easily extended to a force distributed over a unit area so that:

$$\sigma = \frac{\sigma_{max} \pi}{2L} x \qquad (1.2.2c)$$

where σ_{max} is the "tensile strength" of the material and has the units of pressure.

If L_o is the equilibrium distance, then the strain ε for a given displacement x is defined as:

$$\varepsilon = \frac{x}{L_o} \qquad (1.2.2d)$$

Thus:

$$\frac{\sigma}{\varepsilon} = \left[\frac{L_o \pi \sigma_{max}}{2L}\right] = E \qquad (1.2.2e)$$

All the terms in the square brackets may be considered constant for any one particular material (for small displacements around the equilibrium position) and can thus be represented by a single property E, the "elastic modulus" or "Young's modulus" of the material. Equation 1.2.2e is a familiar form of Hooke's law, which, in words, states that stress is proportional to strain.

In practice, no material is as strong as the "theoretical" tensile strength. Usually, weaknesses occur due to slippage across crystallographic planes, impurities, and mechanical defects. When stress is applied, fracture usually initiates at these points of weakness, and failure occurs well below the theoretical tensile strength. Values for actual tensile strength in engineering handbooks are obtained from experimental results on standard specimens and so provide a basis for engineering structural design. As will be seen, additional knowledge regarding the geometrical shape and condition of the material is required to determine whether or not fracture will occur in a particular specimen for a given applied stress.

1.2.3 Strain energy

In one dimension, the application of a force F resulting in a small deflection, dx, of an atom from its equilibrium position causes a change in its potential energy, dW. The total potential energy can be determined from Hooke's law in the following manner:

$$dW = Fdx$$
$$F = kx \qquad (1.2.3a)$$
$$W = \int kxdx = \frac{1}{2}kx^2$$

This potential energy, W, is termed "strain energy." Placing a material under stress involves the transfer of energy from some external source into strain potential energy within the material. If the stress is removed, then the strain energy is released. Released strain energy may be converted into kinetic energy, sound, light, or, as shall be shown, new surfaces within the material.

If the stress is increased until the bond is broken, then the strain energy becomes available as bond potential energy (neglecting any dissipative losses due to heat, sound, etc.). The resulting two separated atoms have the potential to form bonds with other atoms. The atoms, now separated from each other, can be considered to be a "surface." Thus, for a solid consisting of many atoms, the atoms on the surface have a higher energy state compared to those in the interior. Energy of this type can only be described in terms of quantum physics. This energy is equivalent to the "surface energy" of the material.

1.2.4 Surface energy

Consider an atom "A" deep within a solid or liquid, as shown in Fig. 1.2.2. Long-range chemical attractive forces and short-range Coulomb repulsive forces act equally in all directions on a particular atom, and the atom takes up an equilibrium position within the material. Now consider an atom "B" on the surface. Such an atom is attracted by the many atoms just beneath the surface *as well as* those further beneath the surface because the attractive forces between atoms are "long-range", extending over many atomic dimensions. However, the corresponding repulsive force can only be supplied by a few atoms just beneath the surface because this force is "short-range" and extends only to within the order of an atomic diameter. Hence, for an equilibrium of forces on a surface atom, the repulsive force due to atoms just beneath the surface must be increased over that which would normally occur. This increase is brought about by movement of the surface atoms inward and thus closer toward atoms just beneath the surface. The closer the surface atoms move toward those beneath the surface, the larger the repulsive force (see Fig. 1.2.1). Thus, atoms on the surface move inward until the repulsive short-range forces from atoms just beneath the surface balance the long-range attractive forces from atoms just beneath and well below the surface.

The surface of the solid or liquid appears to be acting like a thin tensile skin, which is shrink-wrapped onto the body of the material. In liquids, this effect manifests itself as the familiar phenomenon of surface tension and is a consequence of the potential energy of the surface layer of atoms. Surfaces of solids also have surface potential energy, but the effects of surface tension are not readily observable because solids are not so easily deformed as liquids. The surface energy of a material represents the potential that a surface has for making chemical bonds with other like atoms. The surface potential energy is stored as an increase in compressive strain energy within the bonds between the surface atoms and those just beneath the surface. This compressive strain energy arises due to the slight increase in the short-range repulsive force needed to balance the long-range attractions from beneath the surface.

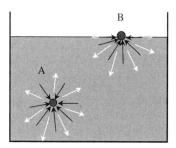

Fig. 1.2.2 Long-range attractive forces and short-range repulsive forces acting on an atom or molecules within a liquid or solid. Atom "B" on the surface must move closer to atoms just beneath the surface so that the resulting short-range repulsive force balances the long-range attractions from atoms just beneath and further beneath the surface.

1.2.5 Stress

Stress in an engineering context means the number obtained when force is divided by the surface area of application of the force. Tension and compression are both "normal" stresses and occur when the force acts perpendicular to the plane under consideration. In contrast, shear stress occurs when the force acts along, or parallel to, the plane. To facilitate the distinction between different types of stress, the symbol σ denotes a normal stress and the symbol τ shear stress. The total state of stress at any point within the material should be given in terms of both normal and shear stresses.

To illustrate the idea of stress, consider an elemental volume as shown in Fig. 1.2.3 (a). Force components dF_x, dF_y, dF_z act normal to the faces of the element in the x, y, and z directions, respectively. The definition of stress, being force divided by area, allows us to express the different stress components using the subscripts i and j, where i refers to the direction of the normal to the plane under consideration and j refers to the direction of the applied force. For the component of force dF_x acting perpendicular to the plane dydz, the stress is a normal stress (i.e., tension or compression):

$$\sigma_{xx} = \frac{dF_x}{dydz} \qquad (1.2.5a)$$

The symbol σ_{xx} denotes a normal stress associated with a plane whose normal is in the x direction (first subscript), the direction of which is also in the x direction (second subscript), as shown in Fig. 1.2.4.

Tensile stresses are generally defined to be positive and compressive stresses negative. This assignment of sign is purely arbitrary, for example, in rock mechanics literature, compressive stresses so dominate the observed modes of failure that, for convenience, they are taken to be positive quantities. The force component dF_y also acts across the dydz plane, but the line of action of the force to the plane is such that it produces a shear stress denoted by τ_{xy}, where, as before, the first subscript indicates the direction of the normal to the plane under consideration, and the second subscript indicates the direction of the applied force. Thus:

$$\tau_{xy} = \frac{dF_y}{dydz} \qquad (1.2.5b)$$

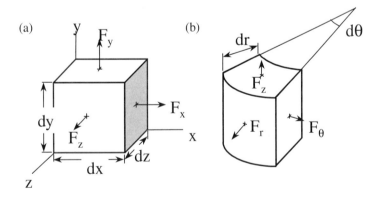

Fig. 1.2.3 Forces acting on the faces of a volume element in (a) Cartesian coordinates and (b) cylindrical-polar coordinates.

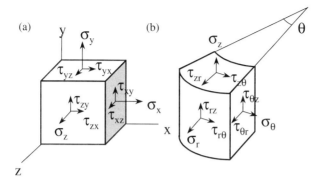

Fig. 1.2.4 Stresses resulting from forces acting on the faces of a volume element in (a) Cartesian coordinates and (b) cylindrical-polar coordinates. Note that stresses are labeled with subscripts. The first subscript indicates the direction of the normal to the plane over which the force is applied. The second subscript indicates the direction of the force. "Normal" forces act normal to the plane, whereas "shear" stresses act parallel to the plane.

For the stress component dF_z acting across $dydz$, the shear stress is:

$$\tau_{xz} = \frac{dF_z}{dydz} \qquad (1.2.5c)$$

Shear stresses may also be assigned direction. Again, the assignment is purely arbitrary, but it is generally agreed that a positive shear stress results when the direction of the line of action of the forces producing the stress and the direction of the outward normal to the surface of the solid are of the same sign; thus, the shear stresses τ_{xy} and τ_{xz} shown in Fig. 1.2.4 are positive. Similar considerations for force components acting on planes $dxdz$ and $dxdy$ yield a total of nine expressions for stress on the element $dxdydz$, which in matrix notation becomes:

$$\begin{bmatrix} \sigma_{xx} & \tau_{xy} & \tau_{xz} \\ \tau_{yx} & \sigma_{yy} & \tau_{yz} \\ \tau_{zx} & \tau_{zy} & \sigma_{zz} \end{bmatrix} \qquad (1.2.5d)$$

The diagonal members of this matrix σ_{ij} are normal stresses. Shear stresses are given by τ_{ij}. If one considers the equilibrium state of the elemental area, it can be seen that the matrix of Eq. 1.2.5d must be symmetrical such that $\tau_{xy} = \tau_{yx}$, $\tau_{yz} = \tau_{zy}$, $\tau_{zx} = \tau_{xz}$. It is often convenient to omit the second subscript for normal stresses such that $\sigma_x = \sigma_{xx}$ and so on.

The nine components of the stress matrix in Eq. 1.2.5d are referred to as the stress tensor. Now, a scalar field (e.g., temperature) is represented by a single value, which is a function of x, y, z:

$$T = f(x, y, z)$$
$$U = [T] \quad (1.2.5e)$$

A vector field (e.g., the electric field) is represented by three components, E_x, E_y, E_z, where each of these components may be a function of position x, y, z*.

$$E = G(E_x, E_y, E_z)$$
$$E = \begin{bmatrix} E_x \\ E_y \\ E_z \end{bmatrix} \quad (1.2.5f)$$

where
$$E_x = f(x, y, z); E_y = g(x, y, z); E_z = h(x, y, z).$$

A tensor field, such as the stress tensor, consists of nine components, each of which is a function of x, y, and z and is shown in Eq. 1.2.5d. The tensor nature of stress arises from the ability of a material to support shear. Any applied force generally produces both "normal" (i.e., tensile and compressive) stresses and shear stresses. For a material that cannot support any shear stress (e.g., a nonviscous liquid), the stress tensor becomes "diagonal." In such a liquid, the normal components are equal, and the resulting "pressure" is distributed equally in all directions.

It is sometimes convenient to consider the total stress as the sum of the average, or mean, stress and the stress *deviations*.

$$\begin{bmatrix} \sigma_x & \tau_{xy} & \tau_{xz} \\ \tau_{yx} & \sigma_y & \tau_{yz} \\ \tau_{zx} & \tau_{zy} & \sigma_z \end{bmatrix} = \begin{bmatrix} \sigma_m & 0 & 0 \\ 0 & \sigma_m & 0 \\ 0 & 0 & \sigma_m \end{bmatrix} + \begin{bmatrix} \sigma_x - \sigma_m & \tau_{xy} & \tau_{xz} \\ \tau_{yx} & \sigma_y - \sigma_m & \tau_{yz} \\ \tau_{zx} & \tau_{zy} & \sigma_z - \sigma_m \end{bmatrix}$$

$$(1.2.5g)$$

The mean stress is defined as:

$$\sigma_m = \frac{1}{3}(\sigma_x + \sigma_y + \sigma_z) \quad (1.2.5h)$$

where it will be remembered that $\sigma_x = \sigma_{xx}$, etc. The remaining stresses, the deviatoric stress components, together with the mean stress, describe the actual

---

* The stress tensor is written with two indices. Vectors require only one index and may be called tensors of the first rank. The stress tensor is of rank 2. Scalars are tensors of rank zero.

state of stress within the material. The mean stress is thus associated with the change in *volume* of the specimen (dilatation), and the deviatoric component is responsible for any change in *shape*. Similar considerations apply to axis-symmetric systems, as shown in Fig. 1.2.3b.

Let us now consider the stress acting on a plane da, which is tilted at an angle θ to the x axis, as shown in Fig. 1.2.5, but whose normal is perpendicular to the z axis.

It can be shown that the normal stress acting on da is:

$$\sigma_\theta = \sigma_x \cos^2\theta + \sigma_y \sin^2\theta + 2\tau_{xy}\sin\theta\cos\theta$$
$$= \frac{1}{2}(\sigma_x + \sigma_y) + \frac{1}{2}(\sigma_x - \sigma_y)\cos 2\theta + \tau_{xy}\sin 2\theta$$

(1.2.5i)

and the shear stress across the plane is found from:

$$\tau_\theta = (\sigma_x - \sigma_y)\sin\theta\cos\theta + \tau_{xy}(\sin^2\theta - \cos^2\theta)$$
$$= \frac{1}{2}(\sigma_x - \sigma_y)\sin 2\theta - \tau_{xy}\cos 2\theta$$

(1.2.5j)

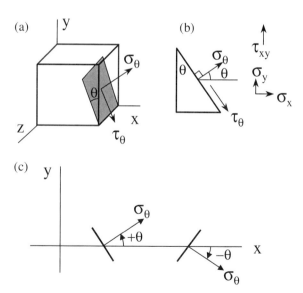

Fig. 1.2.5 (a) Stresses acting on a plane, which makes an angle with an axis. Normal and shear stresses for an arbitrary plane may be calculated using Eqs. 1.2.5i and 1.2.5j. (b) direction of stresses. (c) direction of angles.

From Eq. 1.2.5i, it can be seen that when $\theta = 0$, $\sigma_\theta = \sigma_x$ as expected. Further, when $\theta = \pi/2$, $\sigma_\theta = \sigma_y$. As θ varies from 0 to 360°, the stresses σ_θ and τ_θ vary also and go through minima and maxima. At this point, it is of passing interest to determine the angle θ such that $\tau_\theta = 0$. From Eq. 1.2.5j, we have:

$$\tan 2\theta = \frac{2\tau_{xy}}{\sigma_x - \sigma_y} \qquad (1.2.5k)$$

which, as will be shown in Section 1.2.10, gives the angle at which σ_θ is a maximum.

1.2.6 Strain

1.2.6.1 Cartesian coordinate system

Strain is a measure of relative extension of the specimen due to the action of the applied stress and is given in general terms by Eq. 1.2.2d. With respect to an x, y, z Cartesian coordinate axis system, as shown in Fig. 1.2.6 (a), a point within the solid undergoes displacements u_x, u_y, and u_z and unit elongations, or strains, are defined as[1]:

$$\varepsilon_x = \frac{\partial u_x}{\partial x}; \varepsilon_y = \frac{\partial u_y}{\partial y}; \varepsilon_z = \frac{\partial u_z}{\partial z} \qquad (1.2.6.1a)$$

Normal strains ε_i are positive where there is an extension (tension) and negative for a contraction (compression). For a uniform bar of length L, the change of length as a result of an applied tension or compression may be denoted ΔL. Points within the bar would have a displacement in the x direction that varied according to their distance from the fixed end of the bar. Thus, a plot of displacement u_x vs x would be linear, indicating that the strain ($\partial u_x/\partial x$) is a constant. Thus, at the end of the bar, at x = L, the displacement $u_x = \Delta L$ and thus the strain is $\Delta L/L$.

Shear strains represent the distortion of a volume element. Consider the displacements u_x and u_y associated with the movement of a point P from P_1 to P_2 as shown in Fig. 1.2.7 (a). Now, the displacement u_y increases linearly with x along the top surface of the volume element. Thus, just as we may find the displacement of a particle in the y direction from the normal strain $u_y = \varepsilon_y y$, and since $u_y = (\delta u_y/\delta x)x$, we may define the shear strain $\varepsilon_{xy} = \partial u_y/\partial x$. Similar arguments apply for displacements and shear strains in the x direction.

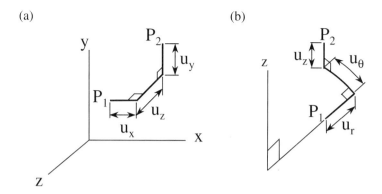

Fig. 1.2.6 Points within a material undergo displacements (a) u_x, u_y, u_z in Cartesian coordinates and (b) u_r, u_θ, u_z in cylindrical polar coordinates as a result of applied stresses.

However, consider the case in Fig. 1.2.7 (b), where $\partial u_y/\partial x$ is equal and opposite in magnitude to $\partial u_x/\partial y$. Here, the volume element has been rotated but not deformed. It would be incorrect to say that there were shear strains given by $\varepsilon_{xy} = -\partial u_y/\partial x$ and $\varepsilon_{yx} = \partial u_x/\partial y$, since this would imply the existence of some strain potential energy in an undeformed element. Thus, it is physically more appropriate to define the shear strain as:

$$\varepsilon_{xy} = \frac{1}{2}\left(\frac{\partial u_x}{\partial y} + \frac{\partial u_y}{\partial x}\right)$$

$$\varepsilon_{yz} = \frac{1}{2}\left(\frac{\partial u_y}{\partial z} + \frac{\partial u_z}{\partial y}\right) \quad (1.2.6.1b)$$

$$\varepsilon_{xz} = \frac{1}{2}\left(\frac{\partial u_x}{\partial z} + \frac{\partial u_z}{\partial x}\right)$$

where it is evident that shearing strains reduce to zero for pure rotations but have the correct magnitude for shear deformations of the volume element.

Many engineering texts prefer to use the angle of deformation as the basis of a definition for shear strain. Consider the angle θ in Fig. 1.2.7 (a). After deformation, the angle θ, initially 90°, has now been reduced by a factor equal to $\partial u_y/\partial x + \partial u_x/\partial y$. This quantity is called the shearing angle and is given by γ_{ij}^1. Thus:

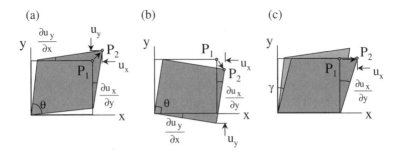

Fig. 1.2.7 Examples of the deformation of an element of material associated with shear strain. A point P moves from P_1 to P_2, leading to displacements in the x and y directions. In (a), the element has been deformed. In (b), the volume of the element has been rotated but not deformed. In (c) both rotation and deformation have occurred.

$$\gamma_{xy} = \frac{\partial u_x}{\partial y} + \frac{\partial u_y}{\partial x}$$

$$\gamma_{yz} = \frac{\partial u_y}{\partial z} + \frac{\partial u_z}{\partial y} \qquad (1.2.6.1c)$$

$$\gamma_{xz} = \frac{\partial u_x}{\partial z} + \frac{\partial u_z}{\partial x}$$

It is evident that $\varepsilon_{ij} = \tfrac{1}{2}\gamma_{ij}$. The symbol γ_{ij} indicates the shearing *angle* defined as the change in angle between planes that were initially orthogonal. The symbol ε_{ij} indicates the shear *strain* component of the strain tensor and includes the effects of rotations of a volume element. Unfortunately, the quantity γ_{ij} is often termed the shear strain rather than the shearing angle since it is often convenient not to carry the factor of 1/2 in many elasticity equations, and in equations to follow, we shall follow this convention.

Figure 1.2.7 (c) shows the situation where both distortion and rotation occur. The degree of distortion of the volume element is the same as that shown in Fig. 1.2.7 (a), but in Fig. 1.2.7 (c), it has been rotated so that the bottom edge coincides with the x axis. Here, $\partial u_y/\partial x = 0$ but the displacement in the x direction is correspondingly greater, and our previous definitions of shear strain still apply. In the special case shown in Fig. 1.2.7 (c), the rotational component of shear strain is equal to the deformation component and is called "simple shear." The term "pure shear" applies to the case where the planes are subjected to shear stresses only and no normal stresses[†]. The shearing angle is positive if there is a

---

[†] An example is the stress that exists through a cross section of a circular bar subjected to a twisting force or torque. In pure shear, there is no change in volume of an element during deformation.

reduction in the shearing angle during deformation and negative if there is an increase.

The general expression for the strain tensor is:

$$\begin{bmatrix} \varepsilon_x & \varepsilon_{xy} & \varepsilon_{xz} \\ \varepsilon_{yx} & \varepsilon_y & \varepsilon_{yz} \\ \varepsilon_{zx} & \varepsilon_{zy} & \varepsilon_z \end{bmatrix} \qquad (1.2.6.1d)$$

and is symmetric since $\varepsilon_{ij} = \varepsilon_{ji}$, etc., and $\gamma_{ij} = 2\varepsilon_{ij}$.

1.2.6.2 Axis-symmetric coordinate system

Many contact stress fields have axial symmetry, and for this reason it is of interest to consider strain in cylindrical-polar coordinates[1,2].

$$\begin{aligned} \varepsilon_r &= \frac{\partial u_r}{\partial r} \\ \varepsilon_\theta &= \frac{u_r}{r} + \frac{1}{r}\frac{\partial u_\theta}{\partial \theta} \\ \varepsilon_z &= \frac{\partial u_z}{\partial z} \end{aligned} \qquad (1.2.6.2a)$$

and for shear "strains"[1,2]:

$$\begin{aligned} \gamma_{rz} &= \frac{\partial u_r}{\partial z} + \frac{\partial u_z}{\partial r} \\ \gamma_{r\theta} &= \frac{\partial u_\theta}{\partial r} - \frac{u_\theta}{r} + \frac{1}{r}\frac{\partial u_r}{\partial \theta} \\ \gamma_{\theta z} &= \frac{1}{r}\frac{\partial u_z}{\partial \theta} + \frac{\partial u_\theta}{\partial z} \end{aligned} \qquad (1.2.6.2b)$$

where u_r, u_θ, and u_z are the displacements of points within the material in the r, θ, and z directions, respectively, as shown in Fig. 1.2.6 (b). Recall also that the shearing angle γ_{ij} differs from the shearing strain ε_{ij} by a factor of 2. In axis-symmetric problems, u_θ is independent of θ, so $\partial u_\theta/\partial \theta = 0$ (also, σ_r and σ_θ are independent of θ and $\tau_{r\theta} = 0$; $\gamma_{r\theta} = 0$); thus, Eq. 1.2.6.2a becomes:

$$\varepsilon_r = \frac{\partial u_r}{\partial r}; \quad \varepsilon_\theta = \frac{u_r}{r}; \quad \varepsilon_z = \frac{\partial u_z}{\partial z} \qquad (1.2.6.2c)$$

Equations 1.2.6.2c are particularly useful for determining the state of stress in indentation stress fields since the displacement of points within the material as a function of r and z may be readily computed (see Chapter 5), and hence the strains and thus the stresses follow from Hooke's law.

1.2.7 Poisson's ratio

Poisson's ratio ν is the ratio of lateral contraction to longitudinal extension, as shown in Fig. 1.2.8. Lateral contractions, perpendicular to an applied longitudinal stress, arise as the material attempts to maintain a constant volume. Poisson's ratio is given by:

$$\nu = \frac{\varepsilon_\perp}{\varepsilon_\parallel} \qquad (1.2.7a)$$

and reaches a maximum value of 0.5, whereupon the material is a fluid, maintains a constant volume (i.e., is incompressible), and cannot sustain shear.

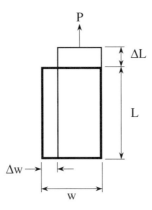

Fig. 1.2.8 The effect of Poisson's ratio is to decrease the width of an object if the applied stress increases its length.

1.2.8 Linear elasticity (generalized Hooke's law)

1.2.8.1 Cartesian coordinate system

In the general case, stress and strain are related by a matrix of constants E_{ijkl} such that:

$$\sigma_{ij} = E_{ijkl}\varepsilon_{kl} \qquad (1.2.8.1a)$$

For an isotropic solid (i.e., one having the same elastic properties in all directions), the constants E_{ijkl} reduce to two, the so-called Lamé constants μ, λ, and can be expressed in terms of two material properties: Poisson's ratio, ν, and Young's modulus, E, where[2]:

$$E = \frac{\mu(3\lambda + 2\mu)}{\lambda + \mu}; \quad \nu = \frac{\lambda}{2(\lambda + \mu)} \qquad (1.2.8.1b)$$

The term "linear elasticity" refers to deformations that show a linear dependence on stress. For applied stresses that result in large deformations, especially in ductile materials, the relationship between stress and strain generally becomes nonlinear.

For a condition of uniaxial tension or compression, Eq. 1.2.2e is sufficient to describe the relationship between stress and strain. However, for the general state of triaxial stresses, one must take into account the strain arising from lateral contraction in determining this relationship. For normal stresses and strains [1,3]:

$$\varepsilon_x = \frac{1}{E}\left[\sigma_x - \nu(\sigma_y + \sigma_z)\right]$$
$$\varepsilon_y = \frac{1}{E}\left[\sigma_y - \nu(\sigma_x + \sigma_z)\right] \qquad (1.2.8.1c)$$
$$\varepsilon_z = \frac{1}{E}\left[\sigma_z - \nu(\sigma_x + \sigma_y)\right]$$

For shear stresses and strains, we have[1,3]:

$$\gamma_{xy} = \frac{1}{G}\tau_{xy}$$
$$\gamma_{yz} = \frac{1}{G}\tau_{yz} \qquad (1.2.8.1d)$$
$$\gamma_{xz} = \frac{1}{G}\tau_{xz}$$

where G is the shear modulus, a high value indicating a larger resistance to shear, given by:

$$G = \frac{E}{2(1+\nu)} \qquad (1.2.8.1e)$$

Also of interest is the bulk modulus K, which is a measure of the compressibility of the material and is found from:

16   Chapter 1. Mechanical Properties of Materials

$$K = \frac{E}{3(1-2\nu)} \quad (1.2.8.1f)$$

1.2.8.2 Axis-symmetric coordinate system

In cylindrical-polar coordinates, Hooke's law becomes[1]:

$$\varepsilon_r = \frac{1}{E}\left[\sigma_r - \nu(\sigma_\theta - \sigma_z)\right]$$
$$\varepsilon_\theta = \frac{1}{E}\left[\sigma_\theta - \nu(\sigma_z - \sigma_r)\right] \quad (1.2.8.2a)$$
$$\varepsilon_z = \frac{1}{E}\left[\sigma_z - \nu(\sigma_r - \sigma_\theta)\right]$$

1.2.9 2-D Plane stress, plane strain

1.2.9.1 States of stress

The state of stress within a solid is dependent on the dimensions of the specimen and the way it is supported. The terms "plane strain" and "plane stress" are commonly used to distinguish between the two modes of behavior for two-dimensional loading systems. In very simple terms, plane strain usually applies to thick specimens and plane stress to thin specimens normal to the direction of applied load.

As shown in Fig. 1.2.9, in plane strain, the *strain* in the thickness, or z direction, is zero, which means that the edges of the solid are fixed or clamped into position; i.e., $u_z = 0$. In plane stress, the *stress* in the thickness direction is zero, meaning that the edges of the solid are free to move. Generally, elastic solutions for plane strain may be converted to plane stress by substituting ν in the solution with $\nu/(1+\nu)$ and plane stress to plane strain by replacing ν with $\nu/(1-\nu)$.

1.2.9.2 2-D Plane stress

In plane stress, Fig. 1.2.9 (a), the stress components in σ_z, τ_{xz}, τ_{yz} are zero and other stresses are uniformly distributed throughout the thickness, or z, direction. Forces are applied parallel to the plane of the specimen, and there are no constraints to displacements on the faces of the specimen in the z direction. Under the action of an applied force, atoms within the solid attempt to find a new equilibrium position by movement in the thickness direction, an amount dependent on the applied stress and Poisson's ratio. Thus, since

$$\sigma_z = 0; \quad \tau_{xz} = 0; \quad \tau_{yz} = 0 \quad (1.2.9.2a)$$

we have from Hooke's law:

$$\varepsilon_z = -\frac{1}{E}\nu(\sigma_x + \sigma_y) \qquad (1.2.9.2b)$$

1.2.9.3 2-D Plane strain

In plane strain, Fig. 1.2.9 (b), it is assumed that the loading along the thickness, or z direction of specimen is uniform and that the ends of the specimen are constrained in the z direction, $u_z = 0$. The resulting stress in the thickness direction σ_z is found from:

$$\sigma_z = \nu(\sigma_x + \sigma_y) \qquad (1.2.9.3a)$$

and also,

$$\varepsilon_z = 0; \quad \tau_{xz} = 0; \quad \tau_{yz} = 0 \qquad (1.2.9.3b)$$

The stress σ_z gives rise to the forces on each end of the specimen which are required to maintain zero net strain in the thickness or z direction. Setting $\varepsilon_z = 0$ in Eq. 1.2.8.1c gives:

$$\frac{\sigma_x}{\varepsilon_x} = \frac{E}{1-\nu^2}$$
$$\frac{\sigma_y}{\varepsilon_y} = \frac{E}{1-\nu^2} \qquad (1.2.9.3c)$$

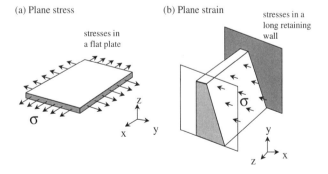

Fig. 1.2.9 Conditions of (a) Plane stress and (b) Plane strain. In plane stress, sides are free to move inward (by a Poisson's ratio effect), and thus strains occur in the thickness direction. In plane strain, the sides of the specimen are fixed so that there are no strains in the thickness direction.

Table 1.2.1 Comparison between formulas for plane stress and plane strain.

	Plane stress	Plane strain
Geometry	Thin	Thick
Normal stresses	$\sigma_z = 0$	$\sigma_z = \nu(\sigma_x + \sigma_y)$ $\sigma_z = \nu(\sigma_r + \sigma_\theta)$
Shear stresses	$\tau_{xz} = 0, \tau_{yz} = 0$	$\tau_{xz} = 0, \tau_{yz} = 0$
Normal strains	$\varepsilon_z = -\frac{1}{E}\nu(\sigma_x + \sigma_y)$	$\varepsilon_z = 0$
Shearing strains	$\gamma_{xz} = 0; \gamma_{yz} = 0$	$\gamma_{xz} = 0, \gamma_{yz} = 0$
Stiffness	E	$E/(1-\nu^2)$

The quantity $E/(1-\nu^2)$ may be thought of as the effective elastic modulus and is greater than the elastic modulus E. The constraint associated with the thickness of the specimen effectively increases its stiffness.

Table 1.2.1 shows the differences in the mathematical expressions for stresses, strains, and elastic modulus for conditions of plane stress and plane strain.

1.2.10 Principal stresses

At any point in a solid, it is possible to find three stresses, σ_1, σ_2, σ_3, which act in a direction normal to three orthogonal planes oriented in such a way that there is no shear stress across those planes. The orientation of these planes of stress may vary from point to point within the solid to satisfy the requirement of zero shear. Only normal stresses act on these planes and they are called the "principal planes of stress." The normal stresses acting on the principal planes are called the "principal stresses." There are no shear stresses acting across the principal planes of stress. The variation in the magnitude of normal stress, at a particular point in a solid, with orientation is given by Eq. 1.2.5i as θ varies from 0 to 360° and shear stress by Eq. 1.2.5j. The stresses σ_θ and τ_θ pass through minima and maxima. The maximum and minimum normal stresses are the principal stresses and occur when the shear stress equals zero. This occurs at the angle indicated by Eq. 1.2.5k. The principal stresses give the maximum normal stress (ie., tension or compression) acting at the point of interest within the solid. The maximum shear stresses act along planes that bisect the principal planes of stress. Since the principal stresses give the maximum values of tensile and compressive stress, they have particular importance in the study of the mechanical strength of solids.

1.2.10.1 Cartesian coordinate system: 2-D Plane stress

The magnitude of the principal stresses for plane stress can be expressed in terms of the stresses that act with respect to planes defined by the x and y axes in a global coordinate system. The maxima and minima can be obtained from the derivative of σ_θ in Eq. 1.2.5i with respect to θ. This yields:

$$\sigma_{1,2} = \frac{\sigma_x + \sigma_y}{2} \pm \sqrt{\left(\frac{\sigma_x - \sigma_y}{2}\right)^2 + \tau_{xy}^2} \qquad (1.2.10.1a)$$

τ_{xy} is the shear stress across a plane perpendicular to the x axis in the direction of the y axis. Since $\tau_{xy} = \tau_{yx}$, then τ_{yx} can also be used in Eq. 1.2.10.1a. σ_1 and σ_2 are the maximum and minimum values of normal stress acting at the point of interest (x,y) within the solid. By convention, the principal stresses are labeled such that $\sigma_1 > \sigma_2$. Note that a very large compressive stress (more negative quantity) may be regarded as σ_2 compared to a very much smaller compressive stress since, numerically, $\sigma_1 > \sigma_2$ by convention. Further confusion arises in the field of rock mechanics, where compressive stresses are routinely assigned positive in magnitude for convenience.

Principal stresses act on planes (i.e., the "principal planes") whose normals are angles θ_p and $\theta_p + \pi/2$ to the x axis as shown in Fig. 1.2.10 (a). Since the stresses σ_1 and σ_2 are "normal" stresses, then the angle θ_p, being the direction of the normal to the plane, also gives the direction of stress. The angle θ_p is calculated from:

$$\tan 2\theta_p = \frac{2\tau_{xy}}{\sigma_x - \sigma_y} \qquad (1.2.10.1b)$$

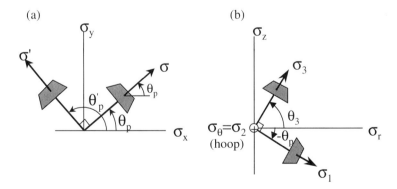

Fig. 1.2.10 Principal planes of stress. (a) In Cartesian coordinates, the principal planes are those whose normals make an angle of θ and θ'_p as shown. In an axis-symmetric state of stress, (b), the hoop stress is always a principal stress. The other principal stresses make an angle of θ_p with the radial direction.

This angle was shown to be that corresponding to a plane of zero shear in Section 1.2.5, Eq. 1.2.5k.

The maximum and minimum values of shearing stress occur across planes oriented midway between the principal planes of stress. The magnitudes of these stresses are equal but have opposite signs, and for convenience, we refer to them simply as the maximum shearing stress. The maximum shearing stress is half the difference between σ_1 and σ_2:

$$\tau_{max} = \pm\sqrt{\left(\frac{\sigma_x - \sigma_y}{2}\right)^2 + \tau_{xy}^2}$$

$$= \pm\frac{1}{2}(\sigma_1 - \sigma_2) \quad (1.2.10.1c)$$

where the plus sign represents the maximum and the minus, the minimum shearing stress. The angle θ_s with which the plane of maximum shear stress is oriented with respect to the global x coordinate axis is found from:

$$\tan 2\theta_s = \frac{\sigma_x - \sigma_y}{2\tau_{xy}} \quad (1.2.10.1d)$$

There are two values of θ_s that satisfy this equation: θ_s and $\theta_s + 90°$ corresponding to τ_{max} and τ_{min}. The angle θ_s is at 45° to θ_p.

The normal stress that acts on the planes of maximum shear stress is given by:

$$\sigma_m = \frac{1}{2}(\sigma_1 + \sigma_2) \quad (1.2.10.1e)$$

which we may call the "mean" stress. On each of the planes of maximum shearing stress, there is a normal stress which, for the two-dimensional case, is equal to the mean stress σ_m. The mean stress is independent of the choice of axes so that:

$$\sigma_m = \frac{1}{2}(\sigma_1 + \sigma_2)$$

$$= \frac{1}{2}(\sigma_x + \sigma_y) \quad (1.2.10.1f)$$

1.2.10.2 Cartesian coordinate system: 2-D Plane strain

For a condition of plane strain, the maximum and minimum principal stresses in the xy plane, σ_1 and σ_2, are given in Eq. 1.2.10.1a. A condition of plane strain refers to a specimen with substantial thickness in the z direction but loaded by forces acting in the x and y directions only. In plane strain problems, an additional stress is set up in the thickness or z direction an amount proportional to Poisson's ratio and is a principal stress. Hence, for plane strain:

$$\sigma_3 = \sigma_z = \nu(\sigma_x + \sigma_y) = \nu(\sigma_1 + \sigma_2) \tag{1.2.10.2a}$$

Although convention generally requires in general that $\sigma_1 > \sigma_2 > \sigma_3$, we usually refer to σ_z as being the third principal stress in plane strain problems regardless of its magnitude; thus in some situations in plane strain, $\sigma_3 > \sigma_2$.

1.2.10.2 Axis-symmetric coordinate system: 2 dimensions

Symmetry of stresses around a single point exists in many engineering problems, and the associated elastic analysis can be simplified greatly by conversion to polar coordinates (r,θ). In a typical polar coordinate system, there exists a radial stress σ_r and a tangential stress σ_θ, and the principal stresses are found from:

$$\sigma_{1,2} = \frac{\sigma_r + \sigma_\theta}{2} \pm \sqrt{\left(\frac{\sigma_r - \sigma_\theta}{2}\right)^2 + \tau_{r\theta}^2} \tag{1.2.10.2a}$$

$$\tau_{max} = \frac{1}{2}(\sigma_1 - \sigma_2) \tag{1.2.10.2b}$$

$$\tan 2\theta_p = \frac{2\tau_{rz}}{(\sigma_r - \sigma_z)} \tag{1.2.10.2c}$$

The shear stress $\tau_{r\theta}$ reduces to zero for the case of axial symmetry, and σ_r and σ_θ are thus principal stresses in this instance.

1.2.10.3 Cartesian coordinate system: 3 dimensions

As noted above, in a three-dimensional solid, there exist three orthogonal planes across which the shear stress is zero. The normal stresses σ_1, σ_2, and σ_3 on these principal planes of stress are called the principal stresses. At a given point within the solid, σ_1 and σ_3 are the maximum and minimum values of normal stress, respectively, and σ_2 has a magnitude intermediate between that of σ_1 and σ_3. The three principal stresses may be found by finding the values of σ such that the determinant

$$\begin{vmatrix} \sigma_x - \sigma & \tau_{yz} & \tau_{zx} \\ \tau_{xy} & \sigma_y - \sigma & \tau_{zy} \\ \tau_{xz} & \tau_{yz} & \sigma_z - \sigma \end{vmatrix} = 0 \tag{1.2.10.3a}$$

Solution of Eq. 1.2.10.3a, a cubic equation in σ, and the three values of σ so obtained are arranged in order such that $\sigma_1 > \sigma_2 > \sigma_3$. Solution of the cubic equation 1.2.3a is somewhat inconvenient in practice, and the principal stresses σ_1, σ_2, and σ_3 may be more conveniently determined from Eq. 1.2.10.1a using σ_x, σ_y, τ_{xy}, and σ_y, σ_z, τ_{yz}, and then σ_x, σ_z, τ_{xz} in turn and selecting the maximum value obtained as σ_1, the minimum as σ_3, and σ_2 is the maximum of the σ_2's calculated for each combination.

The planes of principal shear stress bisect those of the principal planes of stress. The values of shear stress τ for each of these planes are given by:

$$\frac{1}{2}(\sigma_1 - \sigma_3), \frac{1}{2}(\sigma_3 - \sigma_2), \frac{1}{2}(\sigma_2 - \sigma_1) \tag{1.2.10.3a}$$

Note that no attempt has been made to label the stresses given in Eqs. 1.2.10.3a since it is not known a priori which is the greater except that because definition, $\sigma_1 > \sigma_2 > \sigma_3$, the maximum principal shear stress is given by half the difference of σ_1 and σ_3:

$$\tau_{max} = \frac{1}{2}(\sigma_1 - \sigma_3) \tag{1.2.10.3b}$$

The orientation of the planes of maximum shear stress are inclined at ±45° to the first and third principal planes and parallel to the second.

The normal stresses associated with the principal shear stresses are given by:

$$\frac{1}{2}(\sigma_1 + \sigma_3), \frac{1}{2}(\sigma_3 + \sigma_2), \frac{1}{2}(\sigma_2 + \sigma_1) \tag{1.2.10.3c}$$

The mean stress does not depend on the choice of axes, thus:

$$\sigma_m = \frac{1}{3}(\sigma_x + \sigma_y + \sigma_z)$$
$$= \frac{1}{3}(\sigma_1 + \sigma_2 + \sigma_3) \tag{1.2.10.3d}$$

Note that the mean stress σ_m given here is *not* the normal stress which acts on the planes of principal shear stress, as in the two-dimensional case. The mean stress acts on a plane whose direction cosines l, m, n with the principal axes are equal. The shear stress acting across this plane has relevance for the formulation of a criterion for plastic flow within the material.

1.2.10.4 Axis-symmetric coordinate system: 3 dimensions

Axial symmetry exists in many three-dimensional engineering problems, and the associated elastic analysis can be simplified greatly by conversion to cylindrical polar coordinates (r,θ,z). In this case, it is convenient to consider the radial stress σ_r, the axial stress σ_z, and the hoop stress σ_θ. Due to symmetry within the stress field, the hoop stress is always a principal stress, σ_r, σ_θ, and σ_z are independent of θ, and $\tau_{r\theta} = \tau_{\theta z} = 0$. In indentation problems, it is convenient to label the principal stresses such that:

$$\sigma_{1,3} = \frac{\sigma_r + \sigma_z}{2} \pm \sqrt{\left(\frac{(\sigma_r - \sigma_z)}{2}\right)^2 + \tau_{rz}^2} \qquad (1.2.10.4a)$$

$$\sigma_2 = \sigma_\theta \qquad (1.2.10.4b)$$

$$\tau_{max} = \frac{1}{2}[\sigma_1 - \sigma_3] \qquad (1.2.10.4c)$$

Figure 1.2.10 (b) illustrates these stresses. Using these labels, in the indentation stress field we sometimes find that $\sigma_3 > \sigma_2$, in which case the standard convention $\sigma_1 > \sigma_2 > \sigma_3$ is not strictly adhered to. Note that two of the principal stresses, σ_1 and σ_3, lie in the rz plane (with θ a constant). The directions of the principal stresses with respect to the r axis are given by:

$$\tan 2\theta_p = \frac{2\tau_{rz}}{(\sigma_r - \sigma_z)} \qquad (1.2.10.4d)$$

In Eq. 1.2.10.4d, a positive value of θ_p is taken in an anticlockwise direction from the r axis to the line of action of the stress. However, difficulties arise as this angle passes through 45°, and a more consistent value for θ_p is given by Eq. 5.4.2o in Chapter 5. The planes of maximum shear stress bisect the principal planes, and thus:

$$\tan 2\theta_s = \frac{(\sigma_r - \sigma_z)}{2\tau_{rz}} \qquad (1.2.10.4e)$$

1.2.11 Equations of equilibrium and compatibility

1.2.11.1 Cartesian coordinate system

Equations of stress equilibrium and strain compatibility describe the nature of the *variation* in stresses and strains throughout the specimen. These equations have particular relevance for the determination of stresses and strains in systems that cannot be analyzed by a consideration of stress alone (i.e., statically indeterminate systems).

For a specimen whose applied loads are in equilibrium, the state of internal stress must satisfy certain conditions which, in the absence of any body forces (e.g., gravitational or inertial effects), are given by Navier's equations of equilibrium[1,2]:

$$\frac{\partial \sigma_x}{\partial x} + \frac{\partial \tau_{xy}}{\partial y} + \frac{\partial \tau_{xz}}{\partial z} = 0$$

$$\frac{\partial \tau_{yx}}{\partial x} + \frac{\partial \sigma_y}{\partial y} + \frac{\partial \tau_{yz}}{\partial z} = 0 \qquad (1.2.11a)$$

$$\frac{\partial \tau_{zx}}{\partial x} + \frac{\partial \tau_{zy}}{\partial y} + \frac{\partial \sigma_z}{\partial z} = 0$$

Equations 1.2.11a describe the variation of stress from one point to another throughout the solid. Displacements of points within the solid are required to satisfy compatibility conditions which prescribe the variation in displacements throughout the solid and are given by[1,2,3]:

$$\frac{\partial^2 \varepsilon_x}{\partial y^2} + \frac{\partial^2 \varepsilon_y}{\partial x^2} = \frac{\partial^2 \gamma_{xy}}{\partial x \partial y}$$

$$\frac{\partial^2 \varepsilon_y}{\partial z^2} + \frac{\partial^2 \varepsilon_z}{\partial y^2} = \frac{\partial^2 \gamma_{yz}}{\partial y \partial z} \qquad (1.2.11b)$$

$$\frac{\partial^2 \varepsilon_z}{\partial x^2} + \frac{\partial^2 \varepsilon_x}{\partial z^2} = \frac{\partial^2 \gamma_{zx}}{\partial z \partial x}$$

The compatibility relations imply that the displacements within the material vary smoothly throughout the specimen. Solutions to problems in elasticity generally require expressions for stress components which satisfy both equilibrium and compatibility conditions subject to the boundary conditions appropriate to the problem. Formal methods for determining the nature of such expressions that meet these conditions were demonstrated by Airy in 1862.

1.2.11.2 Axis-symmetric coordinate system

Similar considerations apply to axis-symmetric stress systems, where in cylindrical polar coordinates we have (neglecting body forces)[2]:

$$\frac{\partial \sigma_r}{\partial r} + \frac{1}{r}\frac{\partial \tau_{r\theta}}{\partial \theta} + \frac{\partial \tau_{rz}}{\partial z} + \frac{\sigma_r - \sigma_\theta}{r} = 0$$

$$\frac{\partial \tau_{r\theta}}{\partial r} + \frac{1}{r}\frac{\partial \sigma_\theta}{\partial \theta} + \frac{\partial \tau_{\theta z}}{\partial z} + \frac{2\tau_{r\theta}}{r} = 0 \qquad (1.2.11c)$$

$$\frac{\partial \tau_{rz}}{\partial r} + \frac{1}{r}\frac{\partial \tau_{\theta z}}{\partial \theta} + \frac{\partial \sigma_z}{\partial z} + \frac{\tau_{rz}}{r} = 0$$

where $\tau_{r\theta}$ and $\partial/\partial\theta$ terms reduce to zero for symmetry around the z axis.

1.2.12 Saint-Venant's principle

Saint-Venant's principle[4] facilitates the analysis of stresses in engineering structures. The principle states that if the resultant force and moment remain unchanged (i.e., statically equivalent forces), then the stresses, strains and elastic displacements within a specimen far removed from the application of the force are unchanged and independent of the actual type of loading. For example, in indentation or contact problems, the local deformations beneath the indenter depend upon the geometry of the indenter, but the far-field stress distribution is approximately independent of the shape of the indenter.

1.2.13 Hydrostatic stress and stress deviation

For a given volume element of material, the stresses σ_x, σ_y, σ_z, τ_{xy}, τ_{yz}, τ_{zx}, acting on that element may be conveniently resolved into a mean, or average component and the deviatoric components. The mean, or average, stress is found from:

$$\sigma_m = \frac{\sigma_x + \sigma_y + \sigma_z}{3}$$
$$= \frac{\sigma_1 + \sigma_2 + \sigma_3}{3} \qquad (1.2.13a)$$

In Eq. 1.2.13a, σ_m may be considered the "hydrostatic" component of stress, and it should be noted that its value is independent of the choice of axes and is thus called a stress *invariant*. The hydrostatic component of stress may be considered responsible for the uniform compression, or tension, within the specimen. The mean, or hydrostatic, stress acts on a plane whose direction cosines with the principal axes are $l = m = n = 1/3^{1/2}$. This plane is called the "octahedral" plane. The quantity σ_m is sometimes referred to as the octahedral normal stress. The octahedral plane is parallel to the face of an octahedron whose vertices are on the principal axes.

The remaining stress components required to produce the actual state of stress are responsible for the distortion of the element and are known as the deviatoric stresses, or stress deviations.

$$\sigma_{dx} = \sigma_x - \sigma_m$$
$$\sigma_{dy} = \sigma_y - \sigma_m \qquad (1.2.13b)$$
$$\sigma_{dz} = \sigma_z - \sigma_m$$

The deviatoric components of stress are of particular interest since plastic flow, or yielding, generally occurs as a result of distortion of the specimen rather than the application of a uniform hydrostatic stress. The stress deviations do depend on the choice of axes. They must, since the hydrostatic component does not. Hence, the *principal* stress deviations are:

$$\begin{aligned} \sigma_{d1} &= \sigma_1 - \sigma_m \\ \sigma_{d2} &= \sigma_2 - \sigma_m \\ \sigma_{d3} &= \sigma_3 - \sigma_m \end{aligned} \quad (1.2.13c)$$

The maximum difference in stress deviation is given by σ_{d1} minus σ_{d3} which is easily shown to be directly related to the maximum shear stress defined in Eq. 1.2.10.4c.

It is useful to note the following properties associated with the deviatoric components of stress:

$$\begin{aligned} \sigma_o &= \sigma_{d1} + \sigma_{d2} + \sigma_{d3} = \sigma_{dx} + \sigma_{dy} + \sigma_{dz} \\ \frac{\sigma_o^2}{3} &= \frac{1}{2}\left(\sigma_{d1}^2 + \sigma_{d2}^2 + \sigma_{d3}^2\right) \\ &= \frac{1}{6}\left[(\sigma_2 - \sigma_3)^2 + (\sigma_3 - \sigma_1)^2 + (\sigma_1 - \sigma_2)^2\right] \end{aligned} \quad (1.2.13d)$$

where σ_o may be considered a constant that is directly related to the yield stress of the material when this equation is used as a criterion for yield. The shear stress that acts on the octahedral plane is called the "octahedral" shear stress and is given by:

$$\tau_{oct} = \frac{1}{3}\left[(\sigma_2 - \sigma_3)^2 + (\sigma_3 - \sigma_1)^2 + (\sigma_1 - \sigma_2)^2\right]^{1/2} \quad (1.2.13e)$$

1.2.14 Visualizing stresses

It is difficult to display the complete state of stress at points within a material in one representation. It is more convenient to display various attributes of stress on separate diagrams. Stress contours (isobars) are curves of constant stress. Normal or shear stresses may be represented with respect to global, local, or principal coordinate axes. The direction of stress is *not* given by lines drawn normal to the tangents at points on a stress contour. Stress contours give *no* information about the direction of the stress. Stress contours only give information about the magnitude of the stresses.

Stress trajectories, or isostatics, are curves whose tangents show the direction of one of the stresses at the point of tangency and are particularly useful in visualizing the directions in which the stresses act. When stress trajectories are drawn for principal stresses, the trajectories for each of the principal stresses are orthogonal. Tangents to points on stress trajectories indicate the line of action of the stress. Stress trajectories give no information about the magnitude of the stresses at any point.

Some special states of stress are commonly displayed graphically to enable easy comparison with experimental observations. For example, contours obtained by photoelastic methods may be directly compared with shear stress contours. Slip lines occurring in ductile specimens may be compared with shear stress trajectories.

1.3 Plasticity

In many contact loading situations, the elastic limit of the specimen material may be exceeded, leading to irreversible deformation. In the fully plastic state, the material may exhibit strains at a constant applied stress and hence the total strain depends upon the *length of time* the stress, is applied. Thus, we should expect that a theoretical treatment of plasticity involve time rates of change of strain, hence the term "plastic flow."

1.3.1 Equations of plastic flow

Viscosity is resistance to flow. The coefficient of viscosity η is defined such that:

$$\sigma_{zy} = \eta \frac{d\gamma_{zy}}{dt}$$
$$= \eta \frac{d\dot{u}_y}{dz}$$
(1.3.1a)

Equations for fluid flow, where flow occurs at constant volume, are known as the Navier–Stokes equations:

$$\dot{\varepsilon}_x = \frac{1}{3\eta}\left[\sigma_x - \frac{1}{2}(\sigma_y - \sigma_z)\right]$$
$$\dot{\varepsilon}_y = \frac{1}{3\eta}\left[\sigma_y - \frac{1}{2}(\sigma_z - \sigma_x)\right]$$
$$\dot{\varepsilon}_z = \frac{1}{3\eta}\left[\sigma_z - \frac{1}{2}(\sigma_x - \sigma_y)\right]$$
$$\dot{\gamma}_{yz} = \frac{1}{\eta}\sigma_{yz}; \dot{\gamma}_{zx} = \frac{1}{\eta}\sigma_{zx}; \dot{\gamma}_{xy} = \frac{1}{\eta}\sigma_{xy}$$
(1.3.1b)

where $\dot{\gamma}_{xy}$ is the rate of change of shearing strain given by:

28   Chapter 1. Mechanical Properties of Materials

$$\dot{\gamma}_{xy} = \frac{\partial \dot{u}_y}{\partial x} + \frac{\partial \dot{u}_x}{\partial y} \tag{1.3.1c}$$

and so on for yz and zx.

It should be noted that Eqs. 1.3.1b reduce to zero for a condition of hydrostatic stress, indicating that no plastic flow occurs and that it is the deviatoric components of stress that are of particular interest. Thus, Eqs. 1.3.1b can be written:

$$\begin{aligned}
\dot{\varepsilon}_x &= \frac{1}{2\eta}[\sigma_x - \sigma_m] \\
\dot{\varepsilon}_y &= \frac{1}{2\eta}[\sigma_y - \sigma_m] \\
\dot{\varepsilon}_z &= \frac{1}{2\eta}[\sigma_z - \sigma_m] \\
\dot{\gamma}_{yz} &= \frac{1}{\eta}\sigma_{yz}; \dot{\gamma}_{zx} = \frac{1}{\eta}\sigma_{zx}; \dot{\gamma}_{xy} = \frac{1}{\eta}\sigma_{xy}
\end{aligned} \tag{1.3.1d}$$

where σ_m is the mean stress.

Since plastic behavior is so dependent on shear, or deviatoric, stresses, it is convenient to shows stress fields in the plastic regime as "slip-lines." Slip lines are curves whose directions at every point are those of the maximum rate of shear strain at that point. The *maximum* shear stresses occur along two planes that bisect two of the three principal planes, and thus there are two directions of maximum shear strain at each point.

1.4 Stress Failure Criteria

In the previous section, we summarized equations that govern the mechanical behavior of material in the plastic state. Evidently, it is of considerable interest to be able to determine under what conditions a material exhibits elasticity or plasticity. In many cases, plastic flow is considered to be a condition of failure of the specimen under load. Various failure criteria exist that attempt to predict the onset of plastic deformation, and it is not surprising to find that they are concerned with the deviatoric, rather than the hydrostatic, state of stress since it is the former that governs the behavior of the material in the plastic state.

1.4.1 Tresca failure criterion

Shear stresses play such an important role in plastic yielding that Tresca[5] proposed that, in general, plastic deformation occurs when the magnitude of the maximum shear stress τ_{max} reaches half of the yield stress (measured in tension or compression) for the material. A simple example can be seen in the case of uniform tension, where σ_1 equals the applied tensile stress and $\sigma_2 = \sigma_3 = 0$. Yielding will occur when σ_1 reaches the yield stress Y for the material being tested. More generally, the Tresca criterion for plastic flow is:

$$\tau_{max} = \frac{1}{2}(\sigma_1 - \sigma_3)$$
$$= \frac{1}{2}Y \qquad (1.4.1a)$$

or, as is commonly stated:

$$Y = |\sigma_1 - \sigma_3| \qquad (1.4.1b)$$

where σ_1 and σ_3 in these equations are the maximum and minimum principal stresses.

For 2-D plane stress and plane strain, care must be exercised in interpreting and determining the maximum shear stress. Usually, the stress in the thickness direction is labeled σ_3 in these problems, where $\sigma_3 = 0$ for plane stress and $\sigma_3 = \nu(\sigma_1 + \sigma_2)$ for plane strain. In plane strain, the planes of maximum shear stress are usually parallel to the z, or thickness, direction. In plane stress, the maximum shear stress usually occurs across planes at 45° to the z or thickness direction.

1.4.2 Von Mises failure criterion

It is generally observed that the deviatoric, rather than the hydrostatic, component of stress is responsible for failure of a specimen by plastic flow or yielding. In the three-dimensional case, the deviatoric components of stress can be written:

$$\sigma_{d1} = \sigma_1 - \sigma_m$$
$$\sigma_{d2} = \sigma_2 - \sigma_m \qquad (1.4.2a)$$
$$\sigma_{d3} = \sigma_3 - \sigma_m$$

It is desirable that a yield criterion be independent of the choice of axes, and thus we may use the invariant properties of the deviatoric stresses given by Eqs. 1.2.13d, to formulate a useful criterion for plastic flow. According to the von Mises[6] criterion for yield, we have:

$$Y = \sqrt{\frac{1}{2}\left[(\sigma_1 - \sigma_2)^2 + (\sigma_2 - \sigma_3)^2 + (\sigma_3 - \sigma_1)^2\right]} \qquad (1.4.2b)$$

where Y is the yield stress of the material in tension or compression. Equation 1.4.2b can be shown to be related to the strain energy of distortion of the material and is also evidently a description of the octahedral stress as defined by Eq. 1.2.13e. The criterion effectively states that yield occurs when the strain energy of distortion, or the octahedral shear stress, equals a value that is characteristic of the material.

For the special case of plane strain, $\varepsilon_z = 0$, stresses and displacements in the xy plane are independent of the value of z. The z axis corresponds to a principal plane, say $\sigma_z = \sigma_3$. This leads to $\sigma_3 = \frac{1}{2}(\sigma_1+\sigma_2)$ for an incompressible material ($\nu = 0.5$). Equation 1.4.2b can then be written:

$$\tau_{max} = \frac{1}{\sqrt{3}} Y \tag{1.4.2c}$$

where τ_{max} is as given in Eq. 1.2.10.3b.

For the special case where any two of the principal stresses are equal, the Tresca and von Mises criteria are the same. The choice of criterion depends somewhat on the particular application, although the von Mises criterion is more commonly used by the engineering community since it appears to be more in agreement with experimental observations for most materials and loading systems.

The two failure criteria considered above deal with the onset of plastic deformation in terms of shear stresses within the material. In brittle materials, failures generally occur due to the growth of cracks, and only in special applications would one encounter plastic deformations. However, as we shall see in later chapters, plastic deformation of a brittle material routinely occurs in hardness testing where the indentation stress field offers conditions of stress conducive to plastic deformation.

References

1. E. Volterra and J.H. Gaines, *Advanced Strength of Materials*, Prentice–Hall, Englewood Cliffs, N.J., 1971.
2. A.H. Cottrell, *The Mechanical Properties of Matter*, John Wiley & Sons, New York, 1964, p. 135.
3. S.M. Edelglass, *Engineering Materials Science*, Ronald Press Co., New York, 1966.
4. De Saint-Venant, "Mémoire sue l'establissement des équations différentielles des mouvements intérieurs opérés dans les corps solides ductiles au delá des limites où l'élasticité pourrait les ramener à leur premier état," C.R. Bedb. Séances Acad. Sci. Paris, 70, 1870, pp. 474–480.
5. H. Tresca, Mém. Présentées par Divers Savants 18, 1937, p. 733.
6. R. von Mises, Z.Agnew. Math. Mech. 8, 1928, p. 161.

Chapter 2
Linear Elastic Fracture Mechanics

2.1 Introduction

Beginning with the fabrication of stone-age axes, instinct and experience about the strength of various materials (as well as appearance, cost, availability and divine properties) served as the basis for the design of many engineering structures. The industrial revolution of the 19th century led engineers to use iron and steel in place of traditional materials like stone and wood. Unlike stone, iron and steel had the advantage of being strong in tension, which meant that engineering structures could be made lighter and at less cost than was previously possible. In the years leading up to World War 2, engineers usually ensured that the maximum stress within a structure, as calculated using simple beam theory, was limited to a certain percentage of the "tensile strength" of the material. Tensile strength for different materials could be conveniently measured in the laboratory and the results for a variety of materials were made available in standard reference books. Unfortunately, structural design on this basis resulted in many failures because the effect of stress-raising corners and holes on the strength of a particular structure was not appreciated by engineers. These failures led to the emergence of the field of "fracture mechanics." Fracture mechanics attempts to characterize a material's resistance to fracture—its "toughness."

2.2 Stress Concentrations

Progress toward a quantitative definition of toughness began with the work of Inglis[1] in 1913. Inglis showed that the local stresses around a corner or hole in a stressed plate could be many times higher than the average applied stress. The presence of sharp corners, notches, or cracks serves to concentrate the applied stress at these points. Inglis showed, using elasticity theory, that the degree of stress magnification at the edge of the hole in a stressed plate depended on the radius of curvature of the hole.

The smaller the radius of curvature, the greater the stress concentration. Inglis found that the "stress concentration factor", κ, for an elliptical hole is equal to:

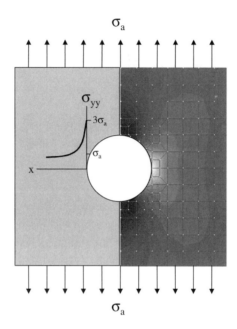

Fig. 2.2.1 Stress concentration around a hole in a uniformly stressed plate. The contours for σ_{yy} shown here were generated using the finite-element method. The stress at the edge of the hole is 3 times the applied uniform stress.

$$\kappa = 1 + 2\sqrt{\frac{c}{\rho}} \qquad (2.2a)$$

where c is the hole radius and ρ is the radius of curvature of the tip of the hole.

For a very narrow elliptical hole, the stress concentration factor may be very much greater than one. For a circular hole, Eq. 2.2a gives $\kappa = 3$ (as shown in Fig. 2.2.1). It should be noted that the stress concentration factor does not depend on the absolute size or length of the hole but only on the *ratio* of the size to the radius of curvature.

2.3 Energy Balance Criterion

In 1920[2], Alan A. Griffith of the Royal Aircraft Establishment in England became interested in the effect of scratches and surface finish on the strength of machine parts subjected to alternating loads. Although Inglis's theory showed that the stress increase at the tip of a crack or flaw depended only on the geometrical shape of the crack and not its absolute size, this seemed contrary to the well-known fact that larger cracks are propagated more easily than smaller ones.

This anomaly led Griffith to a theoretical analysis of fracture based on the point of view of minimum potential energy. Griffith proposed that the reduction in strain energy due to the formation of a crack must be equal to or greater than the increase in surface energy required by the new crack faces. According to Griffith, there are two conditions necessary for crack growth:

i. The bonds at the crack tip must be stressed to the point of failure. The stress at the crack tip is a function of the stress concentration factor, which depends on the ratio of its radius of curvature to its length.
ii. For an increment of crack extension, the amount of strain energy released must be greater than or equal to that required for the surface energy of the two new crack faces.

The second condition may be expressed mathematically as:

$$\frac{dU_s}{dc} \geq \frac{dU_\gamma}{dc} \quad (2.3a)$$

where U_s is the strain energy, U_γ is the surface energy, and dc is the crack length increment. Equation 2.3a says that for a crack to extend, the rate of strain energy release per unit of crack extension must be at least equal to the rate of surface energy requirement. Griffith used Inglis's stress field calculations for a very narrow elliptical crack to show that the strain energy released by introducing a double-ended crack of length 2c in an infinite plate of unit width under a uniformly applied stress σ_a is [2]:

$$U_s = \frac{\pi \sigma_a^2 c^2}{E} \text{ Joules (per meter width)} \quad (2.3b)$$

We can obtain a semiquantitative appreciation of Eq. 2.3b by considering the strain energy released over an area of a circle of diameter 2c, as shown in Fig. 2.3.1. The strain energy is $U = (½\sigma^2/E)(\pi c^2)$. The actual strain energy computed by rigorous means is exactly twice this value as indicated by Eq. 2.3b.

As mentioned in Chapter 1, for cases of plane strain, where the thickness of the specimen is significant, E should be replaced by $E/(1-v^2)$. In this chapter, we omit the $(1-v^2)$ factor for brevity, although it should be noted that in most practical applications it should be included.

The total surface energy for *two* surfaces of unit width and length 2c is:

$$U_\gamma = 4\gamma c \text{ Joules (per meter width)} \quad (2.3c)$$

The factor 4 in Eq. 2.3c arises because of there being two crack surfaces of length 2c. γ is the *fracture* surface energy of the solid. This is usually larger than the surface free energy since the process of fracture involves atoms located a small distance into the solid away from the surface. The fracture surface energy may additionally involve energy dissipative mechanisms such as microcracking, phase transformations, and plastic deformation.

Thus, taking the derivative with respect to c in Eq. 2.3b and 2.3c, this gives us the strain energy release rate (J/m per unit width) and the surface energy creation rate (J/m per unit width). The critical condition for crack growth is:

$$\frac{\pi \sigma_a^2 c}{E} \geq 2\gamma \qquad (2.3d)$$

The left-hand side of Eq. 2.3d is the rate of strain energy release per crack tip and applies to a double-ended crack in an infinite solid loaded with a uniformly applied tensile stress. Equation 2.3d shows that strain energy release rate per increment of crack length is a linear function of crack length and that the required rate of surface energy per increment of crack length is a constant. Equation 2.3d is the Griffith energy balance criterion for crack growth, and the relationships between surface energy, strain energy, and crack length are shown in Fig. 2.3.2.

A crack will not extend until the strain energy release rate becomes equal to the surface energy requirement. Beyond this point, more energy becomes available by the released strain energy than is required by the newly created crack surfaces which leads to unstable crack growth and fracture of the specimen.

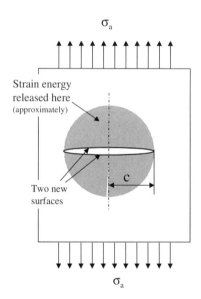

Fig. 2.3.1 The geometry of a straight, double-ended crack of unit width and total length 2c under a uniformly applied stress σ_a. Stress concentration exists at the crack tip. Strain energy is released over an approximately circular area of radius c. Growth of crack creates new surfaces.

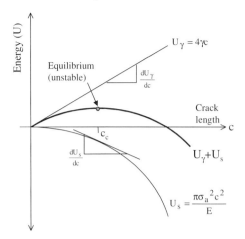

Fig. 2.3.2 Energy versus crack length showing strain energy released and surface energy required as crack length increases for a uniformly applied stress as shown in Fig. 2.3.1. Cracks with length below c_c will not extend spontaneously. Maximum in the total crack energy denotes an unstable equilibrium condition.

The equilibrium condition shown in Fig. 2.3.2 is *unstable*, and fracture of the specimen will occur at the equilibrium condition. The presence of instability is given by the second derivative of Eq. 2.3b. For $d^2U_s/dc^2 < 0$, the equilibrium condition is unstable. For $d^2U_s/dc^2 > 0$, the equilibrium condition is stable. Figure 2.3.3 shows a configuration for which the equilibrium condition is stable. In this case, crack growth occurs at the equilibrium condition, but the crack only extends into the material at the same rate as the wedge.

The energy balance criterion indicates whether crack growth is possible, but whether it will actually occur depends on the state of stress at the crack tip. A crack will not extend until the bonds at the crack tip are loaded to their tensile strength, even if there is sufficient strain energy stored to permit crack growth. For example, if the crack tip is blunted or rounded, then the crack may not extend because of an insufficient stress concentration. The energy balance criterion is a necessary, but not a sufficient condition for fracture. Fracture only occurs when the stress at the crack tip is sufficient to break the bonds there. It is customary to assume the presence of an infinitely sharp crack tip to approximate the worst-case condition. This does not mean, however, that all solids fail upon the immediate application of a load. In practice, stress singularities that arise due to an "infinitely sharp" crack tip are avoided by plastic deformation of the material. However, if such an infinitely sharp crack tip could be obtained, then the crack would not extend unless there was sufficient energy for it to do so.

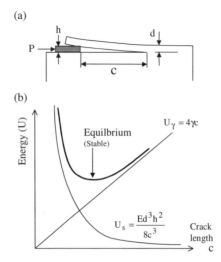

Fig. 2.3.3 (a) Example of stable equilibrium (Obreimoff's experiment). (b) Energy versus crack length showing stable equilibrium as indicated by the minimum in the total crack energy.

For a given stress, there is a minimum crack length that is not self-propagating and is therefore "safe." A crack will not extend if its length is less than the critical crack length, which, for a given uniform stress, is:

$$c_c = \frac{2\gamma E}{\pi \sigma_a^2} \tag{2.3e}$$

In the analyses above, Eq. 2.3b implicitly assumes that the material is linearly elastic and γ in Eq. 2.3d is the fracture surface energy, which is usually greater than the intrinsic surface energy due to energy dissipative mechanisms in the vicinity of the crack tip.

The discussion above refers to a decrease in strain potential energy with increasing crack length. This type of loading would occur in a "fixed-grips" apparatus, where the load is applied, and the apparatus clamped into position. It can be shown that exactly the same arguments apply for a "dead-weight" loading, where the fracture surface energy corresponds to a decrease in potential energy of the loading system. The term "mechanical energy release rate," may be more appropriate than "strain energy release rate" but the latter term is more commonly used.

2.4 Linear Elastic Fracture Mechanics

2.4.1 Stress intensity factor

During the Second World War, George R. Irwin[3] became interested in the fracture of steel armor plating during penetration by ammunition. His experimental work at the U.S. Naval Research Laboratory in Washington, D.C. led, in 1957[4], to a theoretical formulation of fracture that continues to find wide application. Irwin showed that the stress field $\sigma(r,\theta)$ in the vicinity of an infinitely sharp crack tip could be described mathematically by:

$$\sigma_{yy} = \frac{K_1}{\sqrt{2\pi r}} \cos\frac{\theta}{2}\left(1 - \sin\frac{\theta}{2}\sin\frac{3\theta}{2}\right) \qquad (2.4.1a)$$

The first term on the right hand side of Eq. 2.4.1a describes the magnitude of the stress whereas the terms involving θ describe its distribution. K_1 is defined as[*]:

$$K_1 = \sigma_a Y\sqrt{\pi c} \qquad (2.4.1b)$$

The coordinate system for Eqs. 2.4.1a and 2.4.1b is shown in Fig. 2.4.1. In this equation, σ_a is the externally applied stress and c is the crack half-length. K_1 is called the "stress intensity factor." There is an important reason for the stress intensity factor to be defined in this way. For a particular crack system, π and Y are constants so the stress intensity factor tells us that the magnitude of the stress at position (r,θ) depends only on the external stress applied and the square root of the crack length. For example, doubling the externally applied stress σ_a will double the magnitude of the stress in the vicinity of the crack tip at coordinates (r,θ) for a given crack size. Increasing the crack length by 4 times will double the stress at (r,θ) for the same value of applied stress. The stress intensity factor K_1, which includes both applied stress and crack length, is a combined "scale factor," which characterizes the magnitude of the stress at some coordinates (r,θ) near the crack tip. The shape of the stress *distribution* around the crack tip is exactly the same for cracks of all lengths. It is the stress intensity factor that provides information about the *magnitude* of the stresses.

Equation 2.4.1a shows that, for all sizes of cracks, the stresses at the crack tip are infinite. Despite this, the Griffith energy balance criterion must be satisfied for such a crack to extend in the presence of an applied stress σ_a. The stress intensity factor K_1 thus provides a numerical "value," which quantifies the magnitude *of the effect* of the stress singularity at the crack tip. We shall see later that there is a critical value for K_1 for different materials which corresponds to

---

[*] Some authors prefer to define K_1 without $\pi^{1/2}$ in Eq. 2.4.1b. In this case, $\pi^{-1/2}$ does not appear in Eq. 2.4.1a.

the energy balance criterion being met. In this way, this critical value of K_1 characterizes the fracture strength of different materials.

In Eq. 2.4.1b, Y is a function whose value depends on the geometry of the specimen, and σ_a is the applied stress. For a straight double-ended crack in an infinite solid, Y = 1. For a *small* single-ended surface crack (i.e., a semi-infinite solid), Y = $1.12^{5,6}$. This 12% correction arises due to the additional release in strain potential energy (compared with a completely embedded crack) caused by the presence of the free surface near the crack[†] as indicated by the shaded portion in Fig. 2.4.1. This correction has a diminished effect as the crack extends deeper into the material. For embedded penny-shaped cracks, Y = $2/\pi$. For half-penny-shaped surface flaws in a semi-infinite solid, the appropriate value is Y = 0.713. Values of Y for common crack geometries and loading conditions can be found in standard engineering texts.

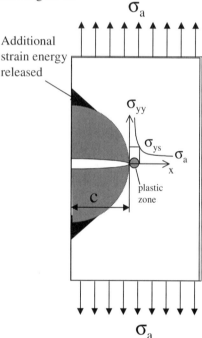

Fig. 2.4.1 Semi-infinite plate under a uniformly applied stress with single-ended surface crack of half-length c. Dark shaded area indicates additional release in strain energy due to the presence of the surface compared to a fully embedded crack in an infinite solid.

[†] A further correction can be made for the effect of a free surface in front of the crack (i.e., the surface to which the crack is approaching). This correction factor is very close to 1 for cracks with a length less than one-tenth the width of the specimen.

Equation 2.4.1a arises from Westergaard's solution[7] for the Airy stress function, which fulfills the equilibrium equations of stresses subject to the boundary conditions associated with a sharp crack, $\rho = 0$, in an infinite, biaxially loaded plate. Equation 2.4.1a applies only to the material in the vicinity of the crack tip. A cursory examination of Eq. 2.4.1a shows that σ_{yy} approaches zero for large values of r rather than the applied stress σ_a. To obtain values for stresses further from the crack tip, additional terms in the series solution must be included. However, near the crack tip, the localized stresses are usually very much greater than the applied uniform stress that may exist elsewhere, and the error is thus negligible.

The subscript 1 in K_1 is associated with tensile loading, as shown in Fig. 2.4.2. Stress intensity factors exist for other types of loading, as also shown in this figure, but our interest centers mainly on type 1 loading—the most common type that leads to brittle failure. An important property of the stress intensity factors is that they are additive for the same type of loading. This means that the stress intensity factor for a complicated system of loads may be derived from the addition of the stress intensity factors determined for each load considered individually.

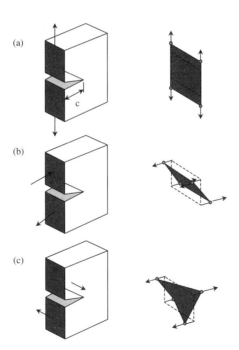

Fig. 2.4.2 Three modes of fracture. (a) Mode I, (b) Mode II, and (c) Mode III. Type I is the most common. The figures on the right indicate displacements of atoms at the crack tip.

We shall show later how the additive property of K_1 permits the stress field in the vicinity of a crack to be calculated on the basis of the stress field that existed in the solid prior to the introduction of the crack.

The power of Eq. 2.4.1b cannot be overestimated. It provides information about events at the crack tip in terms of easily measured macroscopic variables. It implies that the magnitude and distribution of stress in the vicinity of the crack tip can be considered separately and that a criterion for failure need only be concerned with the "magnitude" or "intensity" of stress at the crack tip. Although the stress at an infinitely sharp crack tip may be "infinite" due to the singularity that occurs there, the stress intensity factor is a measure of the "strength" of the singularity.

2.4.2 Crack tip plastic zone

Equation 2.4.1a implies that at $r = 0$ (i.e., at the crack tip) σ_{yy} approaches infinity. However, in practice, the stress at the crack tip is limited to at least the yield strength of the material, and hence linear elasticity cannot be assumed within a certain distance of the crack tip (see Fig. 2.4.1). This nonlinear region is sometimes called the "crack tip plastic zone[8]." Outside the plastic zone, displacements under the externally applied stress mostly follow Hooke's law, and the equations of linear elasticity apply. The elastic material outside the plastic zone transmits stress to the material inside the zone, where nonlinear events occur that may preclude the stress field from being determined exactly. Equation 2.4.1a shows that the stress is proportional to $1/r^{1/2}$. The strain energy release rate is not influenced much by events within the plastic zone *if the plastic zone is relatively small*. It can be shown that an approximate size of the plastic zone is given by:

$$r_p = \frac{K_1^2}{2\pi\sigma_{ys}^2} \qquad (2.4.2a)$$

where σ_{ys} is the yield strength (or yield stress) of the material.

The concept of a plastic zone in the vicinity of the crack tip is one favored by many engineers and materials scientists and has useful implications for fracture in metals. However, the existence of a crack tip plastic zone in brittle solids appears to be objectionable on physical grounds. The stress singularity predicted by Eq. 2.4.1a may be avoided in brittle solids by nonlinear, but elastic, deformations. In Chapter 1, we saw how linear elasticity applies between two atoms for *small displacements* around the equilibrium position. At the crack tip, the displacements are not small on an atomic scale, and nonlinear behavior is to be expected. In brittle solids, strain energy is absorbed by the nonlinear stretching of atomic bonds, not plastic events, such as dislocation movements, that may be expected in a ductile metal. Hence, brittle materials do not fall to pieces under the application of even the smallest of loads even though an infinitely large stress appears to exist at the tip of any surface flaws or cracks within it. The energy balance criterion must be satisfied for such flaws to extend.

2.4.3 Crack resistance

The assumption that all the strain energy is available for surface energy of new crack faces does not apply to ductile solids where other energy dissipative mechanisms exist. For example, in crystalline solids, considerable energy is consumed in the movement of dislocations in the crystal lattice and this may happen at applied stresses well below the ultimate strength of the material. Dislocation *movement* in a ductile material is an indication of yield or plastic deformation, or plastic flow.

Irwin and Orowan[9] modified Griffith's equation to take into account the non-reversible energy mechanisms associated with the plastic zone by simply including this term in the original Griffith equation:

$$\frac{dU_s}{dc} = \frac{dU_\gamma}{dc} + \frac{dU_p}{dc} \qquad (2.4.3a)$$

The right-hand side of Eq. 2.4.3a is given the symbol R and is called the crack resistance. At the point where the Griffith criterion is met, the crack resistance indicates the minimum amount of energy required for crack extension in J/m^2 (i.e., J/m per unit crack width). This energy is called the "work of fracture" (units J/m^2) which is a measure of toughness.

Ductile materials are tougher than brittle materials because they can absorb energy in the plastic zone, as what we might call "plastic strain energy," which is no longer available for surface (i.e., crack) creation. By contrast, brittle materials can only dissipate stored elastic strain energy by surface area creation. The work of fracture is difficult to measure experimentally.

2.4.4 K_{IC}, the critical value of K_I

The stress intensity factor K_I is a "scale factor" which characterizes the magnitude of the stress at some coordinates (r,θ) near the crack tip. If each of two cracks in two different specimens are loaded so that K_I is the same in each specimen, then the magnitude of the stresses in the vicinity of each crack is precisely the same. Now, if the applied stresses are increased, keeping the same value of K_I in each specimen, then eventually the energy balance criterion will be satisfied and the crack in each will extend. The stresses at the crack tip are exactly the same at this point although unknown (theoretically infinite for a perfectly elastic material but limited in practice by inelastic deformations). The value of K_I at the point of crack extension is called the critical value: K_{IC}.

K_{IC} then defines the onset of crack extension. It does not necessarily indicate fracture of the specimen—this depends on the crack stability. It is usually regarded as a material property and can be used to characterize toughness. In contrast to the work of fracture, its determination does not depend on exact knowledge of events within the plastic zone. Consistent and reproducible values of K_{1C} can only be obtained when specimens are tested in plane strain. In plane stress, the critical value of K_1 for fracture depends on the thickness of the plate. Hence, K_{1C} is often called the "plane strain fracture toughness" and has units MPa m$^{1/2}$. Low values of K_{1C} mean that, for a given stress, a material can only withstand a small length of crack before a crack extends.

The condition $K_1 = K_{1C}$ does not necessarily correspond to fracture, or failure, of the specimen. K_{1C} describes the onset of crack extension. Whether this is a stable or unstable condition depends upon the crack system. Catastrophic fracture occurs when the equilibrium condition is unstable. For cracks in brittle materials initiated by contact stresses, the crack may be initially unstable and then become stable due to the sharply diminishing stress field. For example, in Chapter 7, we find that the variation in strain energy release rate (directly related to K_1), the quantity dG/dc, is initially positive and then becomes negative as the crack becomes longer. In terms of stress intensity factor, the crack is stable when $dK_1/dc < 0$ and unstable when $dK_1/dc > 0$. The condition $K_1 = K_{1C}$ for the stable configuration means that the crack is on the point of extension but will not extend unless the applied stress is increased. If this happens, a new stable equilibrium crack length will result. Under these conditions, each increment of crack extension is sufficient to account for the attendant release in strain potential energy. For the unstable configuration, the crack will immediately extend rapidly throughout the specimen and lead to failure. Under these conditions, for each increment of crack extension there is insufficient surface energy to account for the release in strain potential energy.

2.4.5 Equivalence of G and K

Let G be defined as being equal to the strain energy release rate *per crack tip* and given by the left-hand side of Eq. 2.3d, that is, for a double-ended crack within an infinite solid, the rate of release in strain energy per crack tip is:

$$G = \frac{\pi \sigma^2 c}{E} \qquad (2.4.5a)$$

Thus, substituting Eq. 2.4.1b into Eq. 2.4.5a, we have:

$$G = \frac{K_1^2}{E} \qquad (2.4.5b)$$

When $K_1 = K_{1C}$, then G_c becomes the critical value of the rate of release in strain energy for the material which leads to crack extension and possibly fracture of the specimen. The relationship between K_1 and G is significant because it means that the K_{1C} condition is a necessary and sufficient criterion for crack growth since it embodies both the stress and energy balance criteria. The value of K_{1C} describes the stresses (indirectly) at the crack tip *as well as* the strain energy release rate at the onset of crack extension.

It should be remembered that various corrections to K, and hence G, are required for cracks in bodies of finite dimensions. Whatever the correction, the correspondence between G and K is given in Eq. 2.4.5b.

A factor of π sometimes appears in Eq. 2.4.5b depending on the particular definition of K_1 used. Consistent use of π in all these formulae is essential, especially when comparing equations from different sources. Again, we should recognize that Eq. 2.4.5b applies to plane stress conditions. In practice, a condition of plane strain is more usual, in which case one must include the factor $(1-v^2)$ in the numerator.

2.5 Determining Stress Intensity Factors

2.5.1 Measuring stress intensity factors experimentally

Direct application of Griffith's energy balance criterion is seldom practical because of difficulties in determining work of fracture γ. Furthermore, the Griffith criterion is a necessary but not sufficient condition for crack growth. However, stress intensity factors are more easily determined and represent a necessary *and* sufficient condition for crack growth, but in determining the stress intensity factor, Eq. 2.4.1b cannot be used directly because the shape factor Y is not generally known.

As mentioned previously, $Y = 2/\pi$ applies for an embedded penny shaped circular crack of radius c in an infinite plate. Expressions such as this for other types of cracks and loading geometries are available in standard texts. To find the critical value of K_1, it is necessary simply to apply an increasing load P to a prepared specimen, which has a crack of known length c already introduced, and record the load at which the specimen fractures.

Figure 2.5.1 shows a beam specimen loaded so that the side in which a crack has been introduced is placed in tension. Equation 2.5.1 allows the fracture toughness to be calculated from the crack length c and load P at which fracture of the specimen occurs. Note that in practice the length of the beam specimen is made approximately 4 times its height to avoid edge effects.

$$K_1 = \frac{PS}{BW^{3/2}} \left[\begin{array}{c} 2.9\left(\frac{c}{W}\right)^{1/2} - 4.6\left(\frac{c}{W}\right)^{3/2} + \ldots \\ \ldots 21.8\left(\frac{c}{W}\right)^{5/2} - 37.6\left(\frac{c}{W}\right)^{7/2} + 38.7\left(\frac{c}{W}\right)^{9/2} \end{array} \right] \quad (2.5.1)$$

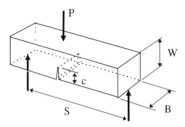

Fig. 2.5.1 Single edge notched beam (SENB)

Consistent and reproducible results for fracture toughness can only be obtained under conditions of plane strain. In plane stress, the values of K_1 at fracture depend on the thickness of the specimen. For this reason, values of K_{1C} are measured in plane strain, hence the term "plane strain fracture toughness."

2.5.2 Calculating stress intensity factors from prior stresses

Under some circumstances, it is possible[10] to calculate the stress intensity factor for a given crack path using the stress field in the solid *before the crack actually exists*. The procedure makes use of the property of superposition of stress intensity factors.

Consider an internal crack of length 2c within an infinite solid, loaded by a uniform externally applied stress σ_a, as shown in Fig. 2.5.2a. The presence of the crack intensifies the stress in the vicinity of the crack tip, and the stress intensity factor K_1 is readily determined from Eq. 2.4.1b. Now, imagine a series of surface tractions in the direction opposite the stress and applied to the crack faces so as to close the crack completely, as shown in Fig. 2.5.2b. At this point, the stress distribution within the solid, uniform or otherwise, is precisely equal to what would have existed in the absence of the crack because the crack is now completely closed. The stress intensity factor thus drops to zero, since there is no longer a concentration of stress at the crack tip. Thus, in one case, the presence of the crack causes the applied stress to be intensified in the vicinity of the crack, and in the other, application of the surface tractions causes this intensification to be reduced to zero.

Consider now the situation illustrated in Fig. 2.5.2c. Wells[11] determined the stress intensity factor K_1 at one of the crack tips A for a symmetric internal crack of total length 2c being loaded by forces F_A applied on the crack faces at a distance b from the center. The value for K_1 for this condition is:

$$K_{1A} = \frac{F_A}{(\pi c)^{1/2}} \left(\frac{c+b}{c-b} \right)^{1/2} \qquad (2.5.2a)$$

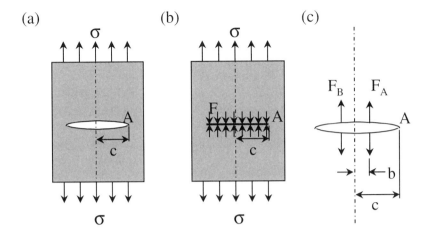

Fig 2.5.2 (a) Internal crack in a solid loaded with an external stress σ. (b) Crack closed by the application of a distribution of surface tractions F. (c) Internal crack loaded with surface tractions F_A and F_B.

Forces F_B also contribute to the stress field at A, and the stress intensity factor due to those forces is:

$$K_{1B} = \frac{F_B}{(\pi c)^{1/2}} \left(\frac{c-b}{c+b}\right)^{1/2} \tag{2.5.2b}$$

Due to the additive nature of stress intensity factors, the total stress intensity factor at crack tip A shown in Fig. 2.5.2c due to forces F_A and F_B, where $F_A = F_B = F$, is‡:

$$K_1 = K_{1A} + K_{1B} = \frac{2F}{\pi^{1/2}} \left(\frac{c}{c^2 - b^2}\right)^{1/2} \tag{2.5.2c}$$

Now, if the tractions F are continuous along the length of the crack, then the force per unit length may be associated with a stress applied σ(b) normal to the crack. The total stress intensity factor is given by integrating Eq. 2.5.2c with F replaced by $dF = \sigma(b)db$.

$$K_1 = \frac{2}{\pi^{1/2}} \int_0^c c^{1/2} \frac{\sigma(b)}{\sqrt{c^2 - b^2}} db \tag{2.5.2d}$$

‡ It is important to note that the Green's weighting functions here apply to a double-ended crack in an infinite solid. For example, Eq. 2.5.2a applies to a force F_A applied to a double-ended symmetric crack and not F_A applied to a single crack tip alone.

However, if the forces F are reversed in sign such that they *close the crack completely*, then the associated stress distribution σ(b) must be that which existed *prior* to the introduction of the crack. The stress intensity factor, as calculated by Eq. 2.5.2d, for continuous surface tractions applied *so as to close the crack*, is precisely the same as that (except for a reversal in sign) calculated for the crack using the macroscopic stress σ_a in the absence of such tractions. For example, for the uniform stress case, where σ(b) = σ_a, Eq. 2.5.2d reduces to Eq. 2.4.1b[§].

As long as the prior stress field within the solid is known, the stress intensity factor for any proposed crack path can be determined using Eq. 2.5.2d. The strain energy release rate G can be calculated from Eq. 2.4.5b. Of course, one cannot always immediately determine whether a crack will follow any particular path within the solid. It may be necessary to calculate strain energy release rates for a number of proposed paths to determine the maximum value for G. The crack extension that results in the maximum value for G is that which an actual crack will follow.

In brittle materials, cracks usually initiate from surface flaws. The strain energy release rate as calculated from the prior stress field (i.e., prior to there being any flaws) applies to the complete growth of the subsequent crack. The conditions determining subsequent crack growth depend on the *prior* stress field. The strain energy release rate, G, can be used to describe the crack growth for all flaws that exist in the prior stress field *but can only be considered applicable for the subsequent growth of the flaw that actually first extends*. Assuming there is a large number of cracks or surface flaws to consider, the one that first extends is that giving the highest value for G (as calculated using the prior stress field) for an increment of crack growth. Subsequent growth of that flaw depends upon the Griffith energy balance criterion (i.e., G ≥ 2γ) being met as calculated along the crack path still using the prior stress field, even though the actual stress field is now different due to the presence of the extending crack.

2.5.3 Determining stress intensity factors using the finite-element method

Stress intensity factors may also be calculated using the finite-element method. The finite-element method is useful for determining the state of stress within a solid where the geometry and loading is such that a simple analytical solution for the stress field is not available. The finite-element solution consists of values for local stresses and displacements at predetermined coordinates called "nodes." A value for the local stress σ_{yy} at a judicious choice of coordinates (r,θ)

[§] To show this, one must make use of the standard integral:

$$\int \frac{1}{\left(a^2 - x^2\right)^{1/2}} dx = \sin^{-1} \frac{x}{a} + C$$

can be used to determine the stress intensity factor K_1. For example, at $\theta = 0$, Eq. 2.4.1a becomes:

$$K_1 = \sigma_{yy}(2\pi r)^{1/2} \qquad (2.5.3a)$$

where σ_{yy} is the magnitude of the local stress at r. It should be noted that the stress at the node that corresponds to the location of the crack tip (r = 0) cannot be used because of the stress singularity there. Stress intensity factors determined for points away from the crack tip, outside the plastic zone, or more correctly the "nonlinear" zone, may only be used. However, one cannot use values that are too far away from the crack tip since Eq. 2.4.1a applies only for small values of r. At large r, σ_{yy} as given by Eq. 2.4.1a approaches zero, and not as is actually the case, σ_a.

Values of K_1 determined from finite-element results and using Eq. 2.5.3a should be the same no matter which node is used for the calculation, subject to the conditions regarding the choice of r mentioned previously. However, it is not always easy to choose which value of r and the associated value of σ_{yy} to use. In a finite-element model, the specimen geometry, density of nodes in the vicinity of the crack tip, and the types of elements used are just some of the things that affect the accuracy of the resultant stress field. One method of optimizing the estimate of K_1 is to determine values for K_1 at different values of r along a line ahead of the crack tip at $\theta = 0$. These values for K_1 are then fitted to a smooth curve and extrapolated to r = 0, as shown in Fig. 2.5.3.

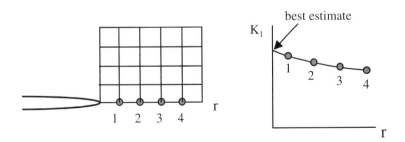

Fig. 2.5.3 Estimating K_1 from finite-element results. For elements near the crack tip, Eq. 2.4.1a is valid and K_1 can be determined from the stresses at any of the nodes near the crack tip. In practice, one needs to determine a range of K_1 for a fixed θ (e.g., $\theta = 0$) for a range of r and extrapolate back to r = 0.

References

1. C.E. Inglis, "Stresses in a plate due to the presence of cracks and sharp corners," Trans. Inst. Nav. Archit. London 55, 1913, pp. 219–230.
2. A.A. Griffith, "Phenomena of rupture and flow in solids," Philos. Trans. R. Soc. London Ser. A221, 1920, pp. 163–198.
3. G.R. Irwin, "Fracture dynamics," Trans. Am. Soc. Met. 40A, 1948, pp. 147–166.
4. G.R. Irwin, "Analysis of stresses and strains near the end of a crack traversing in a plate," J. Appl. Mech. 24, 1957, pp. 361–364.
5. B.R. Lawn, *Fracture of Brittle Solids*, 2nd Ed., Cambridge University Press, Cambridge, U.K., 1993.
6. I.N. Sneddon, "The distribution of stress in the neighbourhood of a crack in an elastic solid," Proc. R. Soc. London, Ser. A187, 1946, pp. 229–260.
7. H.M. Westergaard, "Bearing pressures and cracks," Trans. Am. Soc. Mech. Eng. 61, 1939, pp. A49–A53.
8. D.M. Marsh, "Plastic flow and fracture of glass," Proc. R. Soc. London, Ser. A282, 1964, pp. 33–43.
9. E. Orowan, "Energy criteria of fracture," Weld. J. 34, 1955, pp. 157–160.
10. F.C. Frank and B.R. Lawn, "On the theory of hertzian fracture," Proc. R.. Soc. London, Ser. A229, 1967, pp. 291–306.
11. A.A. Wells, Br. Weld. J. 12, 1965, p. 2.

Chapter 3
Delayed Fracture in Brittle Solids

3.1 Introduction

The fracture of a brittle solid usually occurs due to the growth of a flaw on the surface rather than in the interior. Depending on environmental conditions, brittle solids may exhibit time-delayed failure where fracture may occur some time after the initial application of load. Time-delayed failure of this type usually occurs due to the growth of a pre-existing flaw to the critical size given by the Griffith energy balance criterion. Subcritical crack growth is very important in determining a safe level of operating stress for brittle materials in structural applications. In practice, specimens may be tested for their ability to withstand a design stress for a specified service life by the application of a higher "proof" stress. In this chapter, we investigate the effect of the environment on crack growth in glass, although the general principles apply to other brittle solids. The principles discussed here may be used to determine the service life of a particular specimen subjected to indentation loading where brittle cracking is of concern.

3.2 Static Fatigue

The strength of glass is highly variable and experience shows that it depends on:

i. The rate of loading. Glass is stronger if the load is applied quickly or for short periods. Wiederhorn[1] makes reference to Grenet[2], who in 1899 observed this behavior, but could not account for it. Since then, many other researchers[3-7] have described similar effects.
ii. The degree of abrasion of the surface. A large proportion of fracture mechanics as applied to the strength of brittle solids is devoted to this topic. Work of any significance begins with Inglis in 1913[8] and Griffith in 1920[9].
iii. The humidity of the environment. Orowan[10], in 1944, showed that the surface energy of mica (and hence its fracture toughness) was three and a half times greater in a vacuum than in air that contained a significant proportion of water vapor. Since then, many researchers[11-13] have demon-

strated that the presence of water in conjunction with an applied stress significantly weakens glass.

iv. The temperature. Kropschot and Mikesell[14] in 1957 and other researchers[15-17] showed that the strength of glass increases at low temperatures and that time-dependent fracture is insignificant at cryogenic temperatures.

For most materials, resistance to fracture may be conveniently described by the "plane strain fracture toughness," K_{1C}, introduced in Chapter 2. K_{1C} is the critical value of Irwin's[18] stress intensity factor, K_1, defined as:

$$K_1 = \sigma Y \sqrt{\pi c} \tag{3.2a}$$

where σ is the applied stress, Y is a geometrical shape factor, and c is the crack length. For an applied stress intensity factor $K_1 < K_{1C}$, crack growth may still be possible due to the effect of the environment. Crack growth under these conditions is called "subcritical crack growth" or "static fatigue" and may ultimately lead to fracture some time after the initial application of the load.

Experiments show that there is an applied stress intensity factor $K_1 = K_{1scc}$, which depends on the material, *below* which subcritical crack growth is either undetectable or does not occur at all. K_{1scc} is often called the "static fatigue limit." Experimental results for crack propagation in glass in the vicinity of the static fatigue limit have been widely reported. Shand[7], Wiederhorn and Bolz[19], and Michalske[20] report a fatigue limit for soda-lime glass of 0.25 MPa m$^{1/2}$. Wiederhorn[21] implies a K_{1scc} of 0.3 MPa m$^{1/2}$, and Wan, Latherbai, and Lawn[22] report a static fatigue limit for soda lime glass at about 0.27 MPa m$^{1/2}$. It is generally accepted, however, that more experimental data are needed to clarify whether crack growth ceases entirely for $K_1 < K_{1scc}$ or whether such growth occurs in this domain but at an extremely low rate.

In contrast to the proposed change in crack length described above, Charles[4,5] and Charles and Hillig[23] proposed a mechanism that expresses crack velocity in terms of the thermodynamic and geometrical properties of the crack tip. Charles and Hillig proposed that, depending on the applied stress and the environment, the rate of dissolution of material at the crack tip leads to an increase, a decrease, or no change in the crack tip radius, and hence to corresponding changes in the (Inglis) stress concentration factor over time. The change in stress concentration factor may eventually result in localized stress levels that cause failure of the specimen. Their theory also predicts that, under certain conditions, crack tip blunting leads to a static fatigue limit. It should be noted that Charles and Hillig propose that the change in stress concentration factor is due to the changing geometry of the crack tip, and not to a change in crack length, over time. The stress corrosion theory of Charles and Hillig has considerable historical importance and forms the basis of some present-day architectural glass design strategies. For this reason, it is in our interest to consider it in some detail here.

3.3 The Stress Corrosion Theory of Charles and Hillig

Charles and Hillig[23] developed a theory of time-delayed failure based upon thermodynamic and geometrical considerations. They proposed that the presence of water causes chemical corrosion in glass, which produces a reaction product that is unable to support stress. In addition to being dependent on the chemical potential, the reaction rate also depends on the magnitude of the local stress. The magnitude of the local stress is given by the externally applied stress magnified by the (Inglis) stress concentration factor. Charles and Hillig conjectured that the large stresses at the tip of a flaw or crack cause corrosion to occur preferentially at these sites (see Fig. 3.3.1). This has the effect of changing the crack tip geometry and hence also the local stress level since a change in geometry changes the magnitude of the stress concentration factor. The stress and the corrosion rate at the crack tip are mutually dependent. Charles and Hillig described the velocity of corrosion normal to the interface between the material and the environment by a rate equation of the following form:

$$v = A' \exp\left[\frac{\left(E_o + \gamma_o \frac{V_m}{\rho}\right) - \sigma_1 V^*}{kT}\right] \qquad (3.3a)$$

In this equation, A' is a factor characteristic of the material, E_o is the activation energy in the absence of stress, γ_o is the surface free energy, V_m is molar volume of material, ρ is the radius of curvature of the crack, V^* is defined as the "activation volume" and is equal to the change in activation energy with respect to stress (dE/ds), σ_1 is the local stress at the reaction site—the crack tip, k is Boltzmann's constant, and T is the absolute temperature.

Equation 3.3a gives the crack velocity, or the velocity of the crack front, where it is assumed that the reaction product is incapable of carrying any stress. The magnitude of the stress at the tip of a crack is found from the Inglis stress concentration factor (see Chapter 2), which gives the local stress level expressed in terms of the average applied stress and the crack tip geometry.

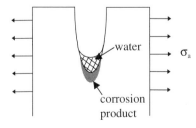

Fig. 3.3.1 Stress corrosion theory of Charles and Hillig.

Chapter 3. Delayed Fracture in Brittle Solids

$$\sigma_l = 2\sigma_a \sqrt{\frac{c}{\rho}} \tag{3.3b}$$

In Eq. 3.3b, ρ is the crack tip radius and a is the crack half length. σ_a is the externally applied tensile stress and σ_l is the local stress at the crack tip.

The crack tip radius can be expressed in terms of its geometry from the second derivative of the displacement of the crack face in the direction parallel to the crack with respect to the displacement in the direction perpendicular to this. Charles and Hillig expressed the time rate of change of the stress concentration factor by a differential equation which, by relating the velocity of the reaction process at the crack face to an analytical expression for the resulting change in the crack tip radius, gives this rate of change in terms of thermodynamic and geometrical parameters. The differential equation has the following form:

$$\frac{d(x/\rho)}{dt} = K\sigma_\ell^n \exp[-\gamma_o/RT] \tag{3.3c}$$

In Eq. 3.3c, K and n are constants, x represents the displacement of the crack boundary into the material, and the other symbols are as in Eq. 3.3a. Charles and Hillig assign the value n = 16 based upon a fit to the experimental results of Mould and Southwick.

Various parameter assignments in Eq. 3.3c lead Charles and Hillig to propose three possible solutions of the rate equation which qualitatively describe experimentally observed events associated with the extension of a flaw under the combined influence of water-induced corrosion and the presence of stress. Charles and Hillig proposed that:

i. The crack may become sharper due to stress corrosion which increases the stress concentration, leading to a corresponding increase in corrosion rate and so on. The crack tip velocity is found from the rate of corrosion of the bulk glass. Fracture eventually occurs after time t_f when the increase in stress concentration results in a tip stress equal to the theoretical strength of the material.

ii. The crack tip may become rounded with the increase in flaw radius balancing the increase in crack length, leading to no increase in the stress concentration factor and no increase in the local stress level. Under these conditions, the crack length increases very slowly, which effectively means that the applied stress can be supported indefinitely.

iii. The crack tip radius and crack width and length may all increase due to corrosion, leading to an effective decrease in the stress concentration factor and hence a decrease in the local stress level and rate of dissolution. Under these conditions, the specimen becomes stronger.

Integration of Eq. 3.3c permits the failure time to be calculated given the applied stress, the temperature, and the geometry of the flaw. Charles and Hillig claim that the rate of change of crack tip radius, rather than the rate of change in crack length, is the parameter most responsible for the change in stress concentration. To simplify the integration, they therefore assume that the crack length remains essentially constant over a limited range of large stresses. This implies that the integration is to be taken over a negligible change in crack length for a large change in tip radius for a given applied stress. This only occurs at applied stresses smaller than would be associated with the fatigue limit, since at the fatigue limit, a small change in stress level results in a change in crack velocity of several orders of magnitude. Combining Eqs. 3.3b and 3.3c, Brown[24] was able to show that for a specific flaw that leads to failure, the integrated form of Eq. 3.3c is:

$$\int_0^{t_f} (\sigma_a/T)^n \exp[-\gamma_0/RT] dt = S \qquad (3.3d)$$

where S is a constant. This important integral forms the basis of modern window glass failure prediction models (e.g., Glass Failure Prediction Model[25,26] and Load Duration Theory[24]).

In Charles and Hillig's theory, item (ii) above implies the existence of a "fatigue limit," which is identified by the condition where the tip stress never reaches the ultimate tensile stress of the material. Charles and Hillig showed that an application of their theory to the experimental data for fatigue strength of glass of Mould and Southwick[6,17] leads to the ratio of the applied stress σ_a at the fatigue limit to the fracture strength σ_n of 0.15 at liquid nitrogen temperatures, where fatigue effects are not in evidence[23]:

$$\frac{\sigma_a}{\sigma_n} = 0.15 \qquad (3.3e)$$

Expressed in terms of (Irwin) stress intensity factors, where K_{Iscc} represents the fatigue limit, this becomes for glass[27]:

$$\begin{aligned} K_{Iscc} &= 0.15 K_{IC} \\ &= 0.117 \text{MPa}\sqrt{m} \end{aligned} \qquad (3.3f)$$

Wiederhorn[21], in 1977, showed that crack velocities fall off sharply when K_I is below 0.3 MPa m$^{1/2}$, a value somewhat larger than that predicted by Charles and Hillig. The existence of the fatigue limit is implicit in Charles and Hillig's work and was given further attention by Marsh[28], who described the fracture and static fatigue of glass in terms of plastic flow at the crack tip. Marsh reported experimental evidence of plastic flow in glass at stresses well below the theoretical tensile strength. The flow stress is dependent on temperature and time. Plastic flow occurs at the highly stressed crack tip and following Irwin, this region is called the "crack tip plastic zone" the size of which is calculated from:

$$r_p = \frac{c\sigma_a^2}{2Y^2} \tag{3.3g}$$

Here, r_p is the radius of the plastic zone, c is the crack length, σ_a is the applied stress, and Y is the yield strength of the material. Failure occurs when the plastic zone reaches a critical size. The "flow" stress σ_y, however, is dependent on time and temperature which permits the fracture stress σ_a to be calculated at different environmental conditions. This work appears to be an alternative statement of Irwin's concept of K_{1C} except that the effects of time and temperature on the variation in fracture strength are included.

Of particular interest is the influence of crack tip plasticity with respect to the static fatigue limit. Marsh stated that Charles and Hillig's theory (which is based upon brittle fracture theory—i.e., Griffith energy balance and Inglis stress concentration factor) requires a crack tip radius of 3 Å when applied to the experimental data of Shand[29] at the lower fatigue limit. Marsh postulated that the selective nature of stress-enhanced corrosion is not likely to result in such small radii. Experimental evidence suggests that the plastic zone should be of order 20 Å and Marsh's calculations show a value for glass of 60 Å. Marsh found that the strength of glass at the static fatigue limit is more accurately described in terms of the size of the plastic zone rather than the rate equation of Charles and Hillig. He claimed that the analysis explains all of the mechanical properties of glass covered by brittle fracture theories and additional issues not covered by them.

It is generally recognized that the stress corrosion theory of Charles and Hillig is insufficient to explain fully the phenomenon of static fatigue although it does attempt to do so in terms of the physics of the phenomenon.

3.4 Sharp Tip Crack Growth Model

Charles and Hillig's theory attempts to describe the physical mechanisms of subcritical crack growth. By contrast, many researchers[32] make no attempt to do so other than to assume a sharp crack tip and to describe subcritical crack velocities using an empirical mathematical model. Crack velocities in Region I of Fig. 3.4.1 are thought to be dependent on both the applied stress intensity factor and the partial pressure of water vapor in the environment. For a static applied stress σ_a, the crack velocity is given by:

$$\frac{dc}{dt} = DK^n \tag{3.4a}$$

In Eq. 3.4a, dc/dt is the crack velocity, and D and n are constants which characterize subcritical crack growth in Region I of Fig. 3.4.1. The time to failure t_f may be found by integrating Eq. 3.4a:

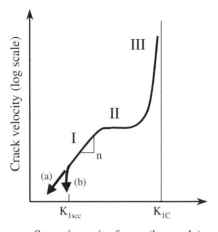

Fig. 3.4.1 Crack velocity versus stress intensity factor showing three regions of crack growth behavior.

$$t_f = \int_{c_i}^{c_f} \frac{1}{D} K_c^{-n} dc \qquad (3.4b)$$

where c_i and c_f are initial and final crack lengths and K_c is the value of K_1 below the value of K_{1C}. t_f is the time for the crack to grow from c_i to c_f.

Equation 3.4a can be written in terms of K_1 using Eq. 3.2a:

$$t_f = \frac{2}{DY^2 \sigma_a^2 \pi} \int_{K_i}^{K_t} K_c^{1-n} dK_c \qquad (3.4c)$$

Integrating Eq. 3.4c gives:

$$t_f = \frac{2\left(K_t^{2-n} - K_i^{2-n}\right)}{Y^2 \sigma_a^2 \pi D(n-2)} \qquad (3.4d)$$

where K_t is the value of K_c at the end of Region I and K_i is the initial value of K_c.

The value of n is typically greater than 10, thus:

$$K_c^{2-n} - K_i^{2-n} \approx -K_i^{2-n}$$

$$t_f = \frac{2K_i^{2-n}}{Y^2 \sigma_a^2 \pi D(n-2)} \qquad (3.4e)$$

and since K_i and c_i are related by Eq. 3.2a:

$$t_f = \frac{2c_i^{1-\frac{n}{2}}}{\pi^{\frac{n}{2}}\sigma_a^n D(n-2)Y^n} \tag{3.4f}$$

Equation 3.4f may also be expressed in terms of a "proof stress" σ_p (see Section 3.5) where:

$$K_i = \frac{K_{1C}\sigma_a}{\sigma_p} \tag{3.4g}$$

Substituting into Eq. 3.4e gives:

$$t_f = \frac{2(\sigma_p/\sigma_a)^{n-2}}{D(n-2)\sigma_a^2 Y^2 \pi K_{1C}^{n-2}} \tag{3.4h}$$

The slope of a plot of log t_f vs log σ_a, using experimental results, yields a value for n, and substitution into Eq. 3.4h allows D to be determined. For glass immersed in water, data from Weiderhorn[21] show that n may be taken to be 17 and logD = −102.6. Other values for n have been experimentally determined and are summarized in Table 3.4.1. There are no values for D available directly from the literature, but these may be obtained from reported experimental data simply by reading the coordinates from the graph, using the value for n, and then using Eq. 3.1.2.

The analysis described above is generally referred to as the sharp tip crack growth model since, in contrast to the stress corrosion theory of Charles and Hillig, it describes the growth of a crack at stresses below the critical stress.

Table 3.4.1 Literature values for n, the subcritical crack growth rate constant, for glass immersed in water.

Source	n	Comments
Matthewson[30]	11	
Ritter[31]	13.4	(abraded)
Ritter[14]	13.0	(acid etched)
Ritter and LaPorte[32]	13	(abraded)
Mould and Southwick[6]	12–14	(abraded)
Wiederhorn and Bolz[19]	16.6	
Simmons and Freiman[33]	18.1	

3.5 Using the Sharp Tip Crack Growth Model

The sharp tip crack growth model provides a convenient way to determine whether a specimen will survive its intended lifetime under the design stress. For example, a manufacturer may wish to guarantee that all specimens will survive a specified lifetime under a given load. To do this, the specimens may be subjected to a "proof stress," which will cause all those specimens that will not last the intended lifetime to fail before being put into service.

Consider the diagram shown in Fig. 3.5.1. During the lifetime t_f of the specimen at a steady applied stress σ_a, flaws of length greater than c_i may grow to the critical size and cause the specimen to fracture. Thus, it is desirable to filter out any specimens in a batch that have a flaw of size greater than c_i. The remaining specimens will have flaws of size less than c_i and those flaws may still undergo subcritical crack growth during time t_f but will not reach the critical size a_c during that time. Specimens with a flaw size greater than c_i can be failed immediately before being placed into service by subjecting all specimens to the critical stress σ_p for that flaw size. σ_p, when applied to all specimens, will cause all those specimens containing flaws of a size greater than c_i to fracture. Those specimens that contain flaws all of size below c_i will not fracture. The remaining unbroken specimens can then be expected not to fail at the applied stress during the time t_f. The flaws that they contain may indeed extend during that time but will not reach the critical size for σ_a.

The proof stress σ_p to apply can be calculated from a rearrangement of Eq. 3.4f

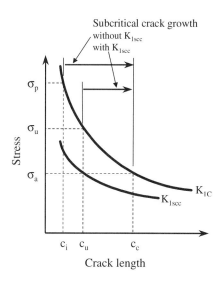

Fig. 3.5.1 Stress versus crack length.

… 58   Chapter 3. Delayed Fracture in Brittle Solids

$$\sigma_p = \left(\frac{t_f \pi \sigma_a^n K_{1C}^{n-2} D(n-2) Y^2}{2}\right)^{\frac{1}{n-2}} \quad (3.5a)$$

The flaw size c_i can be found from:

$$c_i = \left[\frac{t_f \pi^{\frac{n}{2}} \sigma_a^n D(n-2) Y^n}{2}\right]^{\frac{1}{1-\frac{n}{2}}} \quad (3.5b)$$

However, it should be realized that there is strong evidence of the existence of a static fatigue limit K_{1scc}, a stress intensity factor below which subcritical crack growth does not occur. It is possible, depending on the value of σ_a, that the flaw size c_i may be less than that of the flaw size corresponding to the static fatigue limit. Since K_{1scc} is a material property, the critical flaw size for an applied stress σ_a is readily found from:

$$c_u = \left(\frac{K_{1scc}}{\sigma_a Y \sqrt{\pi}}\right)^2 \quad (3.5c)$$

Thus, if c_u happens to be larger than c_i for a given value of σ_a, then only those flaws larger than c_u will undergo subcritical crack growth during the time t_f. In this case, the proof stress required is less than that given by Eq. 3.5a as shown in Fig. 3.5.1. The critical stress for a flaw size c_u can be found from:

$$K_{1C} = \sigma_u \sqrt{\pi c_u} \quad (3.5d)$$

Which proof stress should be applied, σ_p or σ_u?

i.   Calculate a value for c_u using Eq. 3.5c.
ii. Calculate a value for c_i using Eq. 3.5b.
iii. If c_i is larger than c_u, then the proof stress required is σ_p (from Eq.3.5a). If c_i is less than c_u, then the proof stress required is σ_u (from Eq. 3.5d).

Application of the proof stress will guarantee failure of specimens that contain flaws of the critical size to fail. Flaws of such length may be pre-existing or result from subcritical crack growth *during* loading. But, if subcritical crack growth occurs *during unloading*, then the specimen, at the conclusion of the proof test, may contain flaws of length larger than the critical size. Such potentially dangerous flaws thus will be undetected by the proof test procedure. To minimize the possibility of this occurring, the unloading sequence should be performed as quickly as possible.

References

1. S.M. Wiederhorn, "Influence of water vapour on crack propagation in soda-lime glass," J. Am. Ceram. Soc. 50 8, 1967, pp. 407–414.
2. L. Grenet, "Mechanical strength of glass," Bull. Soc. Enc. Ind. Nat. Paris, (Ser. 5) 4, 1899, pp. 838–848.
3. L.V. Black, "Effect of the rate of loading on the breaking strength of glass," Bull. Am. Ceram. Soc. 15 8, 1935, pp. 274–275.
4. R.J. Charles, "Static fatigue of glass I," J. Appl. Phys. 29 11, 1958, pp. 1549–1553.
5. R.J. Charles, "Static fatigue of glass II," J. Appl. Phys. 29 11, 1958, pp. 1554–1560.
6. R.E. Mould and R.D. Southwick, "Strength and static fatigue of abraded glass under controlled ambient conditions: I General concepts and apparatus," J. Am. Ceram. Soc. 42, 1959, pp. 542–547.
7. E.B. Shand, "Fracture velocity and fracture energy of glass in the fatigue range," J. Am. Ceram. Soc. 44 1, 1961, pp. 21–26.
8. C.E. Inglis, "Stresses in a plate due to the presence of cracks and sharp corners," Trans. Inst. Nav. Archit. (London) 55, 1913, pp. 219–230.
9. A.A. Griffith, "Phenomena of rupture and flow in solids," Philos. Trans. R. Soc. London, Ser. A221, 1920, pp. 163–198.
10. E. Orowan, Nature 154, 1944, p. 341.
11. T.C. Baker and F.W. Preston, "Fatigue of glass under static loads," J. Appl. Phys. 17, 1945, pp. 170–178.
12. G.F. Stockdale, F.V. Tooley, and C.W. Ying, "Changes in the tensile strength of glass caused by water immersion treatment," J. Am. Ceram. Soc. 34, 1951, pp. 116–121.
13. F.R.L. Schoening, "On the strength of glass in water vapour," J. Appl. Phys. 31 10, 1960, pp. 1779–1784.
14. R.H. Kropschot and R.P. Mikesell, "Strength and fatigue of glass at very low temperatures," J. Appl. Phys. 28 5, 1957, pp. 610–614.
15. B. Vonnegut and J.G. Glathart, "Effect of water on strength of glass," J. Appl. Phys. 17 12, 1946, pp. 1082–1085.
16. G.O. Jones and W.E.S. Turner, "Influence of temperature on the mechanical strength of glass," J. Soc. Glass Tech. 26, 113, pp. 35–61.
17. R.E. Mould and R.D. Southwick, "Strength and static fatigue of abraded glass under controlled ambient conditions: II Effect of various abrasions and the universal fatigue curve" J. Am. Ceram. Soc. 42, 1959, pp. 582–592.
18. G. Irwin, "Fracture," in *Handbuch der Physik*, Vol.6, Springer-Verlag, Berlin, 1957, p. 551.
19. S.M. Wiederhorn and L.H. Bolz, "Stress corrosion and static fatigue of glass," J. Am. Ceram. Soc. 53, 10 1970, pp. 543–548.
20. T.A. Michalske in *Fracture Mechanics of Ceramics*, Vol. 5, edited by R. C. Bradt, A. G. Evans, D.P.H. Hasselman and F.F. Lange, Plenum Press, New York, 1983.
21. S.M. Wiederhorn, "Dependence of lifetime predictions on the form of the crack propagation equation," Fracture, 3, Canada, 1977, pp. 893–901.
22. K.T. Wan, S. Lathabai, and B.R. Lawn, "Crack velocity functions and thresholds in brittle solids," J. Eur. Ceram. Soc. 6, 1990, pp. 259–268.

23. R.J. Charles, and W.B. Hillig "The kinetics of glass failure," *Symposium on Mechanical Strength of Glass and Ways of Improving It*. Florence, Italy, Sept. 25–29, 1961. Union Scientifique Continentale due Verre, Charleroi, Belgium, 1962, pp. 511–527.
24. W.G. Brown, "A Practicable Formulation for the Strength of Glass and its Special Application to Large Plates," Publication No. NRC 14372, National Research Council of Canada, Ottawa, November, 1974.
25. W.L. Beason, "A Failure Prediction Model for Window Glass," NTIS Accession No. PB81-148421, Institute for Disaster Research, Texas Tech University, Lubbock, Texas, 1980.
26. W.L. Beason and J.R.Morgan, "Glass failure prediction model," Struct. Div. Am. Soc. Ceram. Eng. 110, 1984, pp. 197–212.
27. Note: Davidge quotes Charles and Hillig as calculating this factor to be 0.17.
28. D.M. Marsh, "Plastic flow and fracture of glass," Proc. R. Soc. London, Ser. A282, 1964, pp. 33–43.
29. E.B. Shand, *Glass Engineering Handbook*, 2nd Ed Maple Press, York, PA, 1958.
30. M.J. Matthewson, "An investigation of the statistics of fracture," in *Strength of Inorganic Glass* edited by C.R.Kurkjian, Plenum Press, New York, 1985.
31. J.E. Ritter Jr. "Dynamic fatigue of soda-lime-silica glass" J. Appl. Phys. 40, 1969, pp. 340–344.
32. J.E.Ritter Jr. and R.P. LaPorte, "Effect of test environment on stress-corrosion susceptibility of glass," J. Am. Ceram. Soc. 58, 1975, pp. 265–267.
33. C.J. Simmons and S.W. Freiman, J.Am.Ceram.Soc. 64, 1981, p. 686.

Chapter 4
Statistics of Brittle Fracture

4.1 Introduction

Fractures in brittle solids usually occur due to the existence of surface flaws or cracks in the presence of a tensile stress field according to the Griffith criterion for crack growth.

$$\frac{\pi \sigma^2 c}{E} \geq 2\gamma \tag{4.1a}$$

The left-hand side of Eq. 4.1a describes the release in strain energy and the right hand side gives the surface energy required for the crack to grow.

Griffith's energy balance criterion can also be expressed in terms of Irwin's stress intensity factor:

$$K_1 = \sigma\sqrt{\pi c} \tag{4.1b}$$

Here, σ is the applied stress and c is the crack length. A geometrical shape factor Y is sometimes included in this definition but is not shown here. The critical value of K_1, called K_{1C}, is the fracture toughness of the material. K_{1C} can be regarded as a single-valued material property for most materials. A crack will extend and possibly lead to fracture of the specimen when:

$$K_1 = K_{1C} \tag{4.1c}$$

Weibull statistics are used to predict the existence of a flaw that is capable of causing specimen failure. Weibull statistics[1] have proved useful in a wide variety of situations not necessarily related to the strength of materials. Weibull statistics, when applied to the fracture of brittle solids, refers to *instantaneous* failure at a particular applied stress. However, the effects of subcritical crack growth can be included for the purposes of predicting the expected lifetime of specimens subjected to a tensile stress.

4.2 Basic Statistics

Let X be some random variable that is associated with some event. For example, X might be the number of heads obtained upon two tosses of a coin. Each value of X has a certain probability of occurring, given by:

$$P(X = x) = f(x) \qquad (4.2a)$$

For example, P(X = 1) may give the probability of obtaining one head in two tosses of a coin. f(x) is called a probability function. Figure 4.2.1 shows f(x) for zero, one, or two heads obtained in two tosses of a coin.

A cumulative probability distribution function for the random variable X may be defined as:

$$P(X \le x) = F(x) \qquad (4.2b)$$

where $P(X \le x) = F(x)$ gives the probability that X takes on some value less than or equal to x. For example, Fig. 4.2.2 shows the probability of obtaining at most zero, one, or two heads in two tosses of a coin. The cumulative probability function can be obtained from the probability function by adding the probabilities for all values of X less than x.

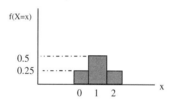

Fig. 4.2.1 Probability function. The y axis gives the probability that the random variable X equals some particular value x, for example, the function shown here indicates the probability of obtaining x number of heads in two tosses of a coin.

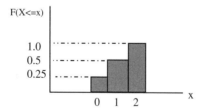

Fig. 4.2.2 Cumulative probability distribution. The y axis gives the probability that the random variable X is equal to or less than a particular value x. The function shown here indicates the probability of obtaining 0, 1, or 2 heads in two tosses of a coin.

Chapter 4. Statistics of Brittle Fracture

$$F(x) = \sum_{u=-\infty}^{x} f(u) \qquad (4.2c)$$

where u in Eq. 4.2c is a dummy variable and takes on all values of x for which u ≤ x. The cumulative distribution function F(x) always increases with increasing values of x.

Now, if the random variable X is a continuous variable, then the probability that X takes on a particular value x is zero. However, the probability that X lies between two different values of x, say a and b, is by definition given by:

$$P(a < x < b) = \int_{a}^{b} f(x)\,dx \qquad (4.2d)$$

where $f(x) \geq 0$ and $\sum_{-\infty}^{+\infty} f(x) = 1$.

Note, that it is the area under the curve of f(x) that gives the probability, as shown in (a) in Fig. 4.2.3. For the continuous case, the value of f(x) at any point is not a probability. Rather, f(x) is called the probability density function.

A cumulative distribution, F(x), for the continuous case gives the probability that X takes on some value ≤ x and can be found from:

$$P(X \leq x) = F(x) = \sum_{-\infty}^{x} f(u) = \int_{-\infty}^{x} f(u)\,du \qquad (4.2e)$$

where u is a dummy variable which takes on all values between minus infinity and x. The value of F(x) approaches 1 with increasing x as shown in Fig. 4.2.3 (b). Equations 4.2d and 4.2e satisfy the basic rules of probability.

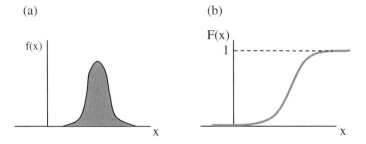

Fig. 4.2.3 (a) Probability density function and (b) cumulative probability distribution function for a continuous random variable X.

4.3 Weibull Statistics

4.3.1 Strength and failure probability

Consider a chain that consists of n links carrying a load W, as shown in Fig. 4.3.1. Because of the load, a stress σ_a is induced in each link of the chain. Let the tensile strength of each link be represented by a continuous random variable S. The value of S may in principle take on all values from $-\infty$ to $+\infty$, but in the present work we may assume that links only fail in tension and hence $S > 0$, or more realistically, $S > \sigma_u$, where $\sigma_u \geq 0$ and is a lower limiting value of tensile strength. All links are said to have a tensile strength equal to or greater than σ_u.

For distributions involving continuous random variables (as in the present case), by definition the chance of any one link having a tensile stress S less than a particular value σ_a is in general given by an integration of the probability density function $f(\sigma)$:

$$F(\sigma) = \int_0^{\sigma_a} f(\sigma) \, d\sigma \qquad (4.3.1a)$$
$$= P(0 < S < \sigma_a)$$

$F(\sigma)$ is the *cumulative* probability function and represents the accumulated area under the probability density function $f(\sigma)$. $F(\sigma)$ increases with increasing σ_a. Since $S > 0$, the total area under $f(\sigma)$ from 0 to $+\infty$ is equal to 1.

If σ_a is an applied stress, what is the probability of failure of the chain? Let the chain have n links. Now, the chain will fail at an applied stress σ_a when any one of the n links has a strength $S \leq \sigma_a$. A larger number of links leads to a greater chance that there exists a weak link in the chain; hence, we expect P_f to increase with n. Let:

Fig. 4.3.1 Chain of n links carrying load W. The chain is only as strong as its weakest link.

$$F(\sigma_a) = P(0 < S < \sigma_a) = \int_0^{\sigma_a} f(\sigma)\,d\sigma \qquad (4.3.1b)$$

where $F(\sigma_a)$ gives the probability of there being a link with $S < \sigma_a$. The probability of there being a link with strength S *greater than* σ_a is:

$$P_S = 1 - F(\sigma_a) \qquad (4.3.1c)$$

because the integral of $f(\sigma)d\sigma$ from zero to infinity equals one.

Thus, the probability that all n links have $S > \sigma_a$ is given by the product of the individual probabilities:

$$\begin{aligned}P_S &= (1-F(\sigma_a)_1)(1-F(\sigma_a)_2)(1-F(\sigma_a)_3)\ldots(1-F(\sigma_a)_n)\\ &= (1-F(\sigma_a))^n\end{aligned} \qquad (4.3.1d)$$

where P_s is the probability of survival for the chain loaded to a stress σ_a and $F(\sigma_a)$ is the same for each link. Equation 4.3.1d gives the probability of the simultaneous nonfailure of all the links.

The probability of failure for the chain is thus:

$$P_f = 1 - (1 - F(\sigma_a))^n \qquad (4.3.1e)$$

It is very important to note that we must express the probability of failure of the chain in terms of the simultaneous probability of nonfailure of all the links. This is because the chain fails when any one of the links has a strength $S \leq \sigma_a$, rather than all the links having $S \leq \sigma_a$. The probability given by Eq. 4.3.1d applies to all n links.

What is $F(\sigma)$? Weibull, for no particular reason other than that of simplicity and convenience, proposed the cumulative probability function:

$$F(\sigma) = 1 - \exp\left[-\left(\frac{\sigma_a - \sigma_u}{\sigma_o}\right)^m\right] \qquad (4.3.1f)$$

where σ_u, σ_o, and m are adjustable parameters, and σ_u represents a stress level below which failure never occurs[*]. As we shall see, σ_o is an indication of the scale of the values of strength and m describes the spread of strengths.

---

[*] There is an alternate three-parameter form which Weibull enunciated and that may be thought to be more academically pleasing than Eq. 4.3.1f. In this alternative form, the probability of failure is given by the difference between the probabilities of failure evaluated at the stress σ and the stress σ_u, adjusted by a factor that represents the total number of flaws are able to cause failure. In this form, we have $F(\sigma) = 1-\exp[-(\sigma_a{}^m - \sigma_u{}^m)/\sigma_o{}^m]$. Sometimes, the parameter σ_u is not included in Eq. 4.3.1f, in which case the equation is referred to as a two-parameter expression.

Substituting Eq. 4.3.1f into 4.3.1e, it is easy to show that the probability of failure for a chain of n links is given by:

$$P_f = 1 - \exp\left[-n\left(\frac{\sigma_a - \sigma_u}{\sigma_o}\right)^m\right] \quad (4.3.1g)$$

Now, this is fine if we know the number of links in advance, however, this may not always be the case, especially when we are dealing with a very large number of links. If ρ is the number or links per unit length, then $n = \rho L$. The probability of failure for the chain P_f may then be computed from:

$$P_f = 1 - \exp\left[-\rho L\left(\frac{\sigma_a - \sigma_u}{\sigma_o}\right)^m\right] \quad (4.3.1h)$$

where L is the total length of the chain, and ρ is the number of links per unit length. The exponent in the Weibull formula is sometimes referred to as the "risk function" and is given the symbol B.

4.3.2 The Weibull parameters

The parameter σ_u represents a lower limit to the tensile strength of each link, where all links have a tensile strength greater than this. The probability of survival for an applied stress $\sigma_a \leq \sigma_u$ is 1.

The parameter m is commonly known as the Weibull modulus and it is the presence of this exponent that provides the statistical basis for the treatment. A high value of m indicates a narrow range in strengths (see Fig. 4.3.2). As m→∞, the *range* of strengths approaches zero, and all links have the same strength.

It is more difficult to give a physical meaning to the parameter σ_o. Various authors give a variety of explanations whereas many do not venture a definition at all. Weibull states "… σ_o is that stress which for the unit of volume gives the probability of rupture S = 0.63."; Davidge[2] gives "…σ_o is a normalizing parameter of no physical significance."; Matthewson[3] says "…σ_o gives the scale of strengths…"; and Atkins and Mai[4] offer: "…a normalizing parameter of no physical significance." σ_o certainly positions the spread of strengths on a scale of tensile strength and for this reason is usually called the "reference strength." However, as we shall see, it does *not* give the position of the maximum number of links with a certain tensile strength in the way that would perhaps be first expected.

The cumulative probability function, $F(\sigma)$, is given by Eq. 4.3.1f. It can be readily shown by integration that the corresponding probability density function, $f(\sigma)$, is, for the case of n = 1, from Eq. 4.3.1a:

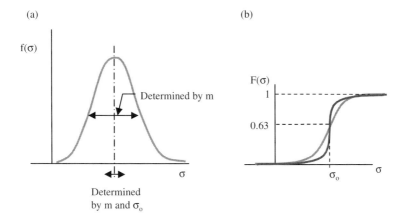

Fig. 4.3.2 Probability density function f(σ) and cumulative probability function F(σ). The effect of values of the Weibull strength parameters is shown.

$$f(\sigma) = \frac{m}{\sigma_o}\left(\frac{\sigma_a - \sigma_u}{\sigma_o}\right)^{m-1} \exp\left[-\left(\frac{\sigma_a - \sigma_u}{\sigma_o}\right)^m\right] \quad (4.3.2a)$$

A plot of f(σ) against σ_a gives a somewhat bell-shaped figure (for m > 1), the width of which depends on m, and the position of which depends on σ_o (see Fig. 4.3.2). For any given value of σ_o, the cumulative probability F(σ), Eq. 4.3.1f, always passes through 0.63 for any value of m. A first derivative test on Eq. 4.3.2a, for the special case of $\sigma_u = 0$, indicates that the maximum value of f(σ) occurs at a stress that is related to σ_o:

$$\sigma_{max} = \sigma_o\left(1 - \frac{1}{m}\right)^{1/m} \quad (4.3.2b)$$

where it is evident that σ_{max} does not equal σ_o (except for m = ∞). Hence, it is evident that σ_o is *not* the stress at which f(σ) rises to a maximum, although it approaches this for large m. In practice, though, m is not particularly large (e.g., for brittle solids, m can be anywhere between 1 and 20) and hence, under these circumstances, the position of σ_o is such that $\sigma_o > \sigma_{max}$ but the difference is not very significant.

The parameter σ_o itself has no real physical significance but indicates the scale of strength. It should be noted that if the applied stress $\sigma_a = \sigma_o$, and for the case of $\sigma_u = 0$, then the probability of failure for each link is 0.63, leading to an undesirably high probability of failure, P_f, for the chain of n links.

The Weibull parameters, m, σ_o, and σ_u may be determined by experiment, and the results so obtained can be used to predict the probability of failure for other specimens of the same surface condition placed under a different stress distribution.

4.4 The Strength of Brittle Solids

4.4.1 Weibull probability function

Consider a brittle solid of area A with area A consisting of a large number of area elements da. The area elements are analogous to the links in the chain in the previous discussion.

  i. Each element da has an associated tensile strength.
 ii. Fracture of the specimen as a result of an applied tensile stress occurs when any one area element fails.
iii. An element fails when it contains a flaw greater than a critical size which depends on the magnitude of the prevailing applied stress (per Griffith).

The probability of failure for an element at a stress σ_a is then related to the probability of that element containing a flaw that is greater than or equal to the critical flaw size.

In general, there may exist flaw distributions in size, density, and orientation on the surface of the solid. The orientation distribution may be combined with size distribution if each flaw that is not normal to the applied stress is given an "equivalent" size as if it were normal. Further, it will be assumed that each flaw that is likely to cause fracture can be assigned an equivalent "penny-shaped" flaw size, a "standard" geometry for fracture analysis.

If ρ is the density of flaws (number per unit area) that could possibly lead to failure for the particular loading condition[†], then the total number of flaws that could lead to failure in the area A is ρA. Later it will be seen that the ρ term (usually unknown) can be conveniently incorporated into the σ_o term (also unknown) to allow a combined parameter to be determined from experimental results.

The Weibull probability function may be expressed:

$$P_f = 1 - \exp\left[-\rho A \left(\frac{\sigma_a - \sigma_u}{\sigma_o}\right)^m\right] \qquad (4.4.1a)$$

[†] It can be seen that the flaw density ρ may be taken as the density of flaws that can conceivably lead to failure. The total probability of failure is given by the product of the individual probabilities of survival as in Eq. 4.3.4. If there are some area elements da that for some reason are incapable of causing failure, then the product $(1-F(\sigma))$ for those elements equals 1 and hence does not contribute to the numerical value of P_s.

In general, the stress may not be uniform over an area A, and thus if σ_a is a function of position, then the following integral is appropriate:

$$P_f = 1 - \exp\left[-\rho \int_0^A \left(\frac{\sigma_a - \sigma_u}{\sigma_o}\right)^m da\right] \quad (4.4.1b)$$

Weibull himself acknowledged that the form of the function $F(\sigma)$ has no theoretical basis but nevertheless serves to give satisfactory results in a large number of practical situations. Since $F(\sigma)$ has three adjustable parameters—m, σ_u, and σ_o—a reasonable fit to experimental data is usually obtainable.

It is customary to incorporate the flaw density term ρ inside the function $F(\sigma)$ so that, for the uniform stress case is:

$$P_f = 1 - \exp\left[-A\left(\frac{\sigma_a - \sigma_u}{\sigma^*}\right)^m\right]$$

where $\quad (4.4.1c)$

$$\sigma^* = \frac{\sigma_o}{\rho^{\frac{1}{m}}}$$

It is evident that ρ and σ_o are interdependent, which is the reason for combining them into a single parameter σ^*. Usually, a value for σ^* can only be determined from suitable fracture experiments. It is very difficult to determine the equivalent, penny-shaped, infinitely sharp, perpendicularly oriented flaw size for every surface flaw on a specimen.

Since σ^* is a property of the surface, it is sometimes useful to write:

$$P_f = 1 - \exp\left[-kA(\sigma_a - \sigma_u)^m\right]$$

where $\quad (4.4.1d)$

$$k = \frac{1}{\sigma^{*m}}$$

which, when $\sigma_u = 0$, becomes:

$$P_f = 1 - \exp\left[-kA\sigma_a^m\right] \quad (4.4.1e)$$

This last expression is a commonly used Weibull probability function and relates the probability of failure for an area A with a surface flaw distribution characterized by m and k subjected to a uniform tensile stress σ_a.

4.4.2 Determining the Weibull parameters

In practice, the Weibull parameters can be found from suitable analysis of experimental data. Rearranging Eq. 4.4.1c gives:

$$\frac{1}{1-P_f} = \exp\left[A\left(\frac{\sigma_a - \sigma_u}{\sigma^*}\right)^m\right] \qquad (4.4.2a)$$

and taking logarithms of both sides twice:

$$\ln\ln\left(\frac{1}{1-P_f}\right) = \ln A + m\ln\left(\frac{\sigma_a - \sigma_u}{\sigma^*}\right) \qquad (4.4.2b)$$

By letting $\sigma_u = 0$ (which is equivalent to saying that there is a probability for failure at every stress level, including zero), then:

$$\ln\ln\left(\frac{1}{1-P_f}\right) = \ln A + m\ln\left(\frac{\sigma_a}{\sigma^*}\right) \qquad (4.4.2c)$$

$$= m\ln\sigma_a + \ln A - m\ln\sigma^*$$

A plot of $\ln\ln(1/(1-P_f))$ vs $\ln \sigma_a$ yields a value for m and σ^*. Any curvature in such a plot implies that σ_u differs from zero. Trial plots for different estimates of σ_u may be made until the most linear curve is obtained. There is no particular reason why strength data should follow the Weibull distribution, and hence a straight line plot may not be possible even with the three adjustable parameters. The only justification for using the technique is that experience has shown that good practical solutions are usually possible.

The probability of failure P_f, for a group of specimens, also gives the ratio of specimens that fail at an applied stress divided by the total number of specimens. To obtain a plot of $\ln\ln(1/(1-P_f))$ vs $\ln \sigma_a$, a large number of specimens, say N, is subjected to a slowly increasing stress σ_a. At convenient intervals of stress, the number of failed specimens is counted (i.e., n). Then, an estimate of the probability of failure at that stress is:

$$P_f = \frac{n}{N} \qquad (4.4.2d)$$

Equation 4.4.2d is called an "estimator." Equation 4.4.2d is not generally used because it is not quite statistically correct. The simplest, most common estimator is:

$$P_f = \frac{n}{N+1} \qquad (4.4.2e)$$

Another common estimator is:

Table 4.4.1 Summary of experimentally determined values of surface flaw parameters m and k.

	As-received glass	Weathered glass
Brown[5,6].	$m = 7.3$ $k = 5.1 \times 10^{-57} \text{ m}^{-2} \text{ Pa}^{-7.3}$ A in sq m, σ in Pa ($k = 5 \times 10^{-30} \text{ sqft}^{-1} \text{ psi}^{-7.3}$, A in sqft, σ psi)	
Beason and Morgan[7].	$m = 9$ $k = 1.32 \times 10^{-69} \text{ m}^{-2}\text{Pa}^{-9}$ ($k = 3.02 \times 10^{-38} \text{ in}^{16} \text{ lb}^{-9}$)	
Beason[8].		$m = 6$ $k = 7.19 \times 10^{-45} \text{ m}^{-2} \text{ Pa}^{-6}$ ($= 4.97 \times 10^{-25} \text{ sq in}^{-1}\text{psi}^{-6}$)

$$P_f = \frac{n - 0.5}{N} \qquad (4.4.2f)$$

The precise form of the estimator is the subject of ongoing research[9]. For example, Eq. 4.4.2e is thought to bias experimental measurements to a lower value for the Weibull modulus.

Table 4.4.1 shows Weibull parameters obtained from various workers for areas of plate window glass. The Weibull parameters determined from experiments using one particular set of samples can in principle be used to predict the probability of failure for other specimens with the same surface condition.

4.4.3 Effect of biaxial stresses

Common sense indicates that a specimen under uniaxial stress will have a lower probability of failure than the same specimen under biaxial stress because in the second case a greater number of flaws will be normal (or nearly so) to an applied tensile stress. So far, we have considered a tensile stress in one direction only acting across an area A. A biaxial, or two-dimensional, stress distribution may be incorporated into the analysis by determining an equivalent one-dimensional stress which acts normal to each flaw.

In the case of biaxial stress, the equivalent stress at some angle to the principal stresses σ_1 and σ_2 can be found, by linear elasticity, from:

$$\sigma_\theta = \left(\sigma_1 \cos^2 \theta + \sigma_2 \sin^2 \theta\right) \qquad (4.4.3a)$$

Chapter 4. Statistics of Brittle Fracture

This then is the equivalent stress which acts normal to a flaw that is oriented at an angle θ to the maximum principal stress. Weibull aimed to reduce the principal stresses to one equivalent stress for each flaw orientation in the specimen. The correction to the risk function B takes the form:

$$B = 2k_1 \int_0^{\pi/2} \int_{-\phi}^{+\phi} \cos^{2m+1}\phi \; \left(\sigma_1 \cos^2\theta + \sigma_2 \sin^2\theta\right)^m d\phi d\theta \qquad (4.4.3b)$$

where ϕ is the angle that the equivalent stress makes with an axis normal to θ and has the range $-\pi/2$ to $+\pi/2$. Equation 4.4.3b is difficult to solve for all but the simplest cases (small m and/or $\sigma_x = \sigma_y$). As an example, Weibull shows that for the case of $\sigma_x = \sigma_y$ and m = 3, the probability of failure is given by:

$$P_f = 1 - \exp\left[-3.2k\sigma^3\right] \qquad (4.4.3c)$$

Weibull's original work actually was based on a one-dimensional tensile stress and applies a correction which increases the probability of failure for the two-dimensional case. The nature of the correction involves an integration of the form (equation 39, Weibull 1939[1]):

$$B = 2k \int_0^{\frac{\pi}{2}} \int_{-\phi}^{+\phi} \left(\sigma_1 \cos^2\theta + \sigma_2 \sin^2\theta\right)^m \cos^{2m+1}\phi \; d\phi d\theta \qquad (4.4.3d)$$

and can only be evaluated readily for small m, or for the case of $\sigma_x = \sigma_y$.

In experimental studies involving flat plates, a biaxial stress distribution exists as a matter of course. Weibull parameters m and k are often determined by experiments involving biaxial stresses, and hence, the biaxial stress correction factor should be applied in a reverse direction. A good example of this procedure is given by Beason[8].

Beason defines C(x,y) as the biaxial stress correction factor to be applied at any particular point on the surface of the plate. At locations where the principal stresses in the two biaxial directions are equal, C(x,y) = 1. σ_{max} is the equivalent principal stress after corrections have been made for time, temperature and humidity as previously described. Beason gives C(x,y) as:

$$C(x,y) = \left[\frac{2}{\pi}\int_0^{\frac{\pi}{2}} \left(\cos^2\theta + n\sin^2\theta\right)^m d\theta\right]^{\frac{1}{m}} \qquad (4.4.3e)$$

where n is the ratio of the minimum to the maximum principal stresses.

Table 4.4.2 Biaxial Stress Correction Factor C(x,y) for m = 7 for different ratios of minimum to maximum principal stress[8].

n	Correction factor
1.0	1.00
0.8	0.92
0.6	0.86
0.4	0.83
0.2	0.81
0.0	0.80
−0.2	0.79
−0.4	0.78
−0.6	0.77
−0.8	0.77
−1.0	0.76

The upper limit of the integration is $\pi/2$ if both principal stresses are tensile. If one is compressive, then the upper limit is given by:

$$\tan^{-1}\left[\left|\frac{\sigma_{max}}{\sigma_{min}}\right|^{\frac{1}{2}}\right] \quad (4.4.3f)$$

The factor C(x,y) decreases as the ratio n increases. Beason and Morgan[7] give a table of values for C(x,y) for ranges of m and n, part of which is reproduced in Table 4.4.2.

Interestingly, it can be shown that if Beason's correction factor is rescaled so that c = 1 at n = 0, then the value of c at n = 1 is very close to that calculated by Weibull. For example, Beason shows that at m = 3 and n = 0:

$$c = \left[\frac{2}{\pi}\int_0^{\frac{\pi}{2}}\left(\cos^2\theta\right)^3 d\theta\right]^{\frac{1}{6}} \quad (4.4.3g)$$

$$= 0.679$$

Now, rescaling so that c = 1 at n = 0, the projected value of c' at n = 1 is:

$$c' = \left(\frac{1}{0.679}\right)^3 \quad (4.4.3h)$$

$$= 3.2$$

74   Chapter 4. Statistics of Brittle Fracture

which is equal to Weibull's correction factor for n = 1 and m = 3. Another example at m = 6 yields c = 0.724. Rescaling so that c = 1 at n = 0, the projected value of c' at n = 1 is 6.91. Weibull's formula (Eq. 4.4.3d) for m = 6, n = 1 yields 4.43. Evidently, Beason's approximation does not hold as well at larger values of m.

4.4.4 Determining the probability of delayed failure

In previous chapters, we have seen that a critical flaw size can be associated with a uniform applied external stress through K_{IC}. This relationship is illustrated in Fig. 3.5.1, where the stress intensity factor is shown in terms of an applied external stress σ_a and crack length c. In this figure, K_{IC} indicates the condition where instantaneous failure occurs and c_c is the critical crack length for a particular value of σ_a.

For an applied stress σ_a, P_f as given by Eq. 4.4.1e is the probability that an area A contains a flaw of size equal to or larger then c_c and is the probability of *instantaneous* failure at that stress. However, if subcritical crack growth occurs during a time t_f, flaws of size c_i, less than c_c, will extend to a length c_c over that time. Thus, for failure *within a time* t_f at stress σ_a we need to know the probability of the area A containing a flaw of size greater than or equal to c_i.

The procedure for determining this probability is very similar to what was seen in Chapter 3 for determining the proof stress σ_p. The proof stress is the critical stress for instantaneous failure for flaws of size c_i, or more correctly, the larger of c_i or c_u, where it will be remembered that c_u is associated with the static fatigue limit. However, instead of calling it a proof stress, we should now think of it as an equivalent applied stress, σ_e. That is, the probability of *instantaneous* failure at a stress σ_e is precisely the same as the probability of *delayed* failure at a stress σ_a. Since the Weibull probability formula gives only the probability of instantaneous failure, we need to use σ_e in Eq. 4.4.1e for determining the probability of delayed failure at an applied stress σ_a. That is:

$$P_f = 1 - \exp\left[-kA\sigma_e^m\right] \qquad (4.4.4a)$$

We must also be aware of the effect of the static fatigue limit. Flaws of size below the static fatigue limit will not undergo subcritical crack growth during the time t_f. Thus, following the same procedure as in Chapter 3, we determine values for c_u and c_i and proceed as follows:

i. Calculate a value for c_u using Eq. 3.5c in Chapter 3.
ii. Calculate a value for c_i using Eq. 3.5b. in Chapter 3.
iii. If c_i is larger than c_u, then the equivalent stress required is σ_p. If c_i is less than c_u, then the equivalent stress required is σ_u. σ_p and σ_u are calculated according to Eqs. 3.5a and 3.5d.

Depending on the magnitude of the applied stress, the static fatigue limit places an upper limit on the probability of failure as calculated by this procedure. For example, for an applied stress of 8 MPa over a 1 m² area, c_i is only greater than c_u for a time to failure less than 60 days[‡]. For longer failure times, c_i is always less than c_u, and the probability of failure approaches a constant value based on the value for the equivalent stress associated with c_u. The time to failure at which P_f approaches a constant value depends upon the applied stress. Table 4.4.3 shows some representative values.

For the situation where $c_u > c_i$, then the equivalent stress becomes:

$$\sigma_e = \frac{K_1}{K_{1scc}} \sigma_a \qquad (4.4.4b)$$

For 4 mm thick, simply supported glass sheets carrying a uniform lateral pressure of 2.2 kPa, the time at which the probability of failure approaches a constant value appears to be about 36 days[§].

The implications of these observations are that failure models are able to predict the probability that an article will fail within the failure time at which P_f approaches a constant value. Designing for longer failure times has no effect on the probability of failure since smaller flaw sizes, which would be predicted to extend to a critical size without considering the static fatigue limit, will not extend because they are below that associated with the static fatigue limit. Thus, if a particular sample lasts longer than this critical time, then as long as the stress level, flaw distribution, and environmental conditions do not change, one would expect the sample to last indefinitely. However, it should be noted that this approach to fracture analysis cannot easily be applied to brittle solids that show an increase in crack resistance with crack extension. For example, a crack in concrete may be arrested by the interface between the cement and a piece of gravel, hence, the failure of the weakest link may not necessarily lead to fracture of the specimen.

Table 4.4.3 Time to failure at which P_f approaches a constant value for some values of applied uniform tensile stress.

Applied stress σ_a (MPa)	Failure time (days) at which P_f is constant
8	60
16	15
22	8

[‡] With other parameters as follows: m=7.3, k=5.1×10^{-57} m^{-2}Pa^{-m}, $\log_{10}D = -102.6$, n=17.

[§] This example corresponds with the recommended lateral pressure for a 1m² area of window glass as specified in various glass design standards.

References

1. W. Weibull, "A statistical theory of the strength of materials," Ingeniorsvetenskapsakademinshandlingar 151, 1939.
2. R.W. Davidge, *Mechanical Behaviour of Ceramics*, Cambridge University Press, Cambridge, U.K., 1979.
3. M.J. Matthewson, "An investigation of the statistics of fracture.," in *Strength of Inorganic Glass* edited by C.R. Kurkjian, Plenum Press, New York, 1985.
4. A.G. Atkins and Y.-W. Mai, *Elastic and Plastic Fracture: Metals, Polymers, Ceramics, Composites, Biological Materials* Ellis Horwood/John Wiley, Chichester, 1985.
5. W.G. Brown, "A Load Duration Theory for Glass Design," National Research Council of Canada, Division of Building Research, NRCC 12354, Ottawa, Ontario, Canada, 1972.
6. W.G. Brown, "A Practicable Formulation for the Strength of Glass and its Special Application to Large Plates," Publication No. NRC 14372, National Research Council of Canada, Ottawa, November 1974.
7. W.L. Beason and J.R. Morgan, "Glass failure prediction model," Struct. Div. Am. Soc. Ceram. Eng. 110 2, 1984, pp. 197–212.
8. W.L. Beason, "A Failure Prediction Model for Window Glass," Institute for Disaster Research, Texas Tech University, Lubbock, Texas, NTIS Accession No. PB81-148421, 1980.
9. J.D. Sullivan and P.H. Lauzon, "Experimental probability estimators for Weibull plots," J. Mater. Sci. Lett. 5, 1986, pp. 1245–1247.

Chapter 5
Elastic Indentation Stress Fields

5.1 Introduction

The nature of the stresses arising from the contact between two elastic bodies is of considerable importance and was first studied by Hertz[1,2] in 1881 before his more well-known work on electricity. Stresses arising from indentations with point loads, spheres, cylindrical flat punches, and diamond pyramids are all of practical interest. The subsequent evolution of the field of contact mechanics has led to applications of the theory to a wide range of disciplines. The *elastic* stress fields generated by an indenter, whether it be a sphere, cylinder, or diamond pyramid, although complex, are well defined. Certain aspects of an indentation stress field, in particular its localized character, make it an ideal tool for investigating the mechanical properties of engineering materials. Before such an investigation can be considered, our first requirement is a detailed knowledge of the elastic stress fields associated with various indenter geometries, and this is the topic of the present chapter. Although a full mathematical derivation of the indentation stress fields associated with a variety of indenters is not given here, enough detail is presented to give an overall picture of how these stresses are calculated from first principles.

5.2 Hertz Contact Pressure Distribution

Hertz was concerned with the nature of the localized deformation and the distribution of pressure between two elastic bodies placed in mutual contact. He sought to assign a shape to the surface of contact that satisfied certain boundary conditions, namely:

i. The displacements and stresses must satisfy the differential equations of equilibrium for elastic bodies, and the stresses must vanish at a great distance from the contact surface.
ii. The bodies are in frictionless contact.
iii. At the surface of the bodies, the normal pressure is zero outside and equal and opposite inside the circle of contact.

iv. The distance between the surfaces of the two bodies is zero inside and greater than zero outside the circle of contact.
v. The integral of the pressure distribution within the circle of contact with respect to the area of the circle of contact gives the force acting between the two bodies.

These conditions define a framework within which a mathematical treatment of the problem may be formulated. Hertz made his analysis general by attributing a quadratic function to represent the profile of the two opposing surfaces and gave particular attention to the case of contacting spheres. Condition 4 above, taken together with the quadric surfaces of the two bodies, defines the form of the contacting surface. Condition 4 notwithstanding, the two contacting bodies are to be considered elastic, semi-infinite, half-spaces. Subsequent elastic analysis is generally based on an appropriate distribution of normal pressure on a semi-infinite half-space, hence our stipulation that, in the formulas to follow, the radius of the circle of contact be very much smaller than the radius of the contacting bodies. By analogy with the theory of electric potential, Hertz deduced that an ellipsoidal distribution of pressure would satisfy the boundary conditions of the problem and found that, for the case of a sphere, the required distribution of pressure is:

$$\frac{\sigma_z}{p_m} = -\frac{3}{2}\left(1 - \frac{r^2}{a^2}\right)^{1/2} \quad r \leq a \tag{5.2a}$$

Hertz did not calculate the magnitudes of the stresses at points throughout the interior but offered a suggestion as to their character by interpolating between those he calculated on the surface and along the axis of symmetry. The stress field associated with indentation of a flat surface with a spherical indenter appears to have been first calculated in detail by Huber[3] in 1904 and again later by Fuchs[4] in 1913, Huber and Fuchs[5] in 1914, and Moreton and Close[6] in 1922. More recently, the integral transform method of Sneddon[7] has been applied to axis-symmetric distributions of normal pressures which correspond to a variety of indenters. An extensive mathematical treatment is given by Gladwell[8], and an accessible text directed to practical applications is that of Johnson[9]. In sections to follow, we summarize some of the most commonly used indentation formulae but without going into their derivation.

5.3 Analysis of Indentation Stress Fields

A mathematical description of the indentation stress field associated with a particular indenter begins with the analysis of the condition of a point contact. This was studied by Boussinesq[10] in 1885. The so-called Boussinesq solution for a point contact allows the stress distribution to be determined for any distribution of pressure within a contact area by the principle of superposition. Any contact configuration, such as indentation with a spherical or cylindrical flat punch indenter, can be viewed as an appropriate distribution of point loads of varying intensity at the specimen surface, and the stress distribution within the interior is given by the superposition of each of the point-load indentation stress fields.

5.3.1 Line contact

The two-dimensional case of a uniformly distributed concentrated force acting along a line, as occurs in a knife edge contact, is of particular interest. The first analytical solution to the problem is attributed to Flamant[11,12]. The distribution of stress within the specimen is radially directed toward the point of contact. At any point r within the specimen, the radial stress, in two-dimensional polar coordinates (see Fig. 5.3.1 for coordinate system), for a load per unit length P perpendicular to the surface of the specimen is given by:

$$\sigma_r = -\frac{2P}{\pi}\frac{\cos\theta}{r} \qquad (5.3.1a)$$
$$\sigma_\theta = \tau_{r\theta} = 0$$

σ_r is a principal stress. The tangential σ_θ and shearing stresses $\tau_{r\theta}$ at any point within the specimen are zero. For any circle of diameter d tangent to the point of application of load and centered on the x axis, it is easy to show that $r = d\cos\theta$, and σ_r given by Eq. 5.3.1a is the same for all points on that circle except for the point r = 0 where a stress singularity occurs (infinite stress and infinite displacement). The stress singularity is avoided in practice by plastic yielding of the specimen material, which serves to spread the load over a small, finite area.

In Cartesian coordinates, the stresses in the xy plane are[13]:

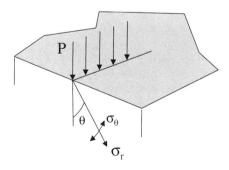

Fig 5.3.1 Polar coordinate system for line contact on a semi-infinite solid.

$$\sigma_x = \frac{2Px^3}{\pi r^4}; \quad \sigma_y = \frac{2Pxy^2}{\pi r^4}; \quad \tau_{xy} = \frac{2Px^2y}{\pi r^4} \qquad (5.3.1b)$$

$$r = \sqrt{x^2 + y^2}$$

5.3.2 Point contact

The stresses within a solid loaded by a point contact were calculated by Boussinesq[10] and are given in cylindrical polar coordinates by Timoshenko and Goodier[12]:

$$\sigma_r = \frac{P}{2\pi}\left[(1-2v)\left[\frac{1}{r^2} - \frac{z}{r^2(r^2+z^2)^{1/2}}\right] - \frac{3r^2z}{(r^2+z^2)^{5/2}}\right]$$

$$\sigma_\theta = \frac{P}{2\pi}(1-2v)\left[-\frac{1}{r^2} + \frac{z}{r^2(r^2+z^2)^{1/2}} + \frac{z}{(r^2+z^2)^{3/2}}\right] \qquad (5.3.2a)$$

$$\sigma_z = -\frac{3P}{2\pi}\frac{z^3}{(r^2+z^2)^{5/2}}$$

$$\tau_{rz} = -\frac{3P}{2\pi}\frac{rz^2}{(r^2+z^2)^{5/2}}$$

Except at the origin, the surface stresses σ_z, τ_{yz}, $\tau_{zx} = 0$. Stresses calculated using Eq. 5.3.2a are shown in Fig. 5.3.2. Note that with Eq. 5.3.2a, and other equations to be presented in later sections, the coordinates r and z are to be entered as positive quantities even though in many texts it is customary to present increasing values of |z| as vertically downward. Also, the load P, customarily shown as acting downward, is also a positive quantity.

The strains corresponding to these stresses may be obtained from Hooke's law, which in cylindrical polar coordinates becomes:

$$\varepsilon_r = \frac{\sigma_r - v(\sigma_\theta + \sigma_z)}{E}$$

$$\varepsilon_\theta = \frac{\sigma_\theta - v(\sigma_r + \sigma_z)}{E} \qquad (5.3.2b)$$

Chapter 5. Elastic Indentation Stress Fields 81

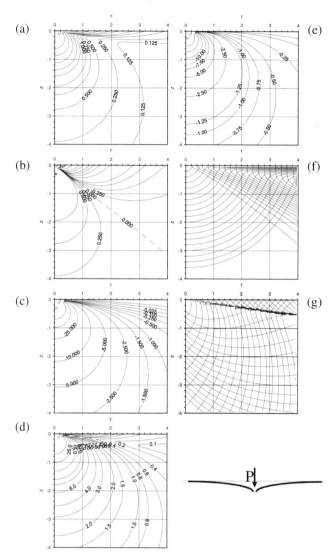

Fig 5.3.2 Stress trajectories and contours of equal stress in MPa for Boussinesq "point load" configuration calculated for load P = 100N and Poisson's ratio $\nu = 0.26$. Distances r and z in mm. (a) σ_1, (b) σ_2, (c) σ_3, (d) τ_{max}, (e) the hydrostatic stress σ_H, (f) σ_1 and σ_3 trajectories, (g) τ_{max} trajectories.

The strains ε_r and ε_θ are related to the displacements by:

$$\varepsilon_r = \frac{\partial u_r}{\partial r}$$

$$\varepsilon_\theta = \frac{u_r}{r}$$

(5.3.2c)

At the surface, $z = 0$, the displacements are:

$$u_r = -\frac{P}{2\pi Er}(1-2v)(1+v)$$

$$u_z = -\frac{P}{\pi Er}(1-v^2)$$

(5.3.2d)

The displacements may be expressed in spherical polar coordinates thus:

$$u_r = \frac{P}{4\pi Gr}\left(4(1-v)\cos\theta - (1-2v)\right)$$

$$u_\theta = \frac{P}{4\pi Gr}\left(\frac{(1-2v)\sin\theta}{1+\cos\theta} - (3-4v)\sin\theta\right)$$

(5.3.2e)

where G is the shear modulus. Note that Eqs. 5.3.2d indicate that u_r and $u_z \to 0$ as $r \to \infty$, thus allowing the displacements to be given with reference to what may be considered "fixed" points, or points on the surface of the specimen at a relatively large distance from the point of contact.

5.3.3 Analysis of stress and deformation

If the contact pressure distribution is known, then the surface deflections and stresses may be obtained by a superposition of those arising from individual point contacts. Consider a general point on the surface G with coordinates (r,θ), as shown in Fig. 5.3.3.

We define a local coordinate system at G by radial and angular variables (s,ϕ). At some local distance s from this point, a pressure dp acts on a small elemental area. The corresponding point force dP is given by:

$$dP = p(s,\phi)sdsd\phi.$$ (5.3.3a)

The deflection of the surface at G due to the point force dP is given by u_z in Eq. 5.3.2d with the variable r being replaced by s. The total deflection of the surface at G is the sum of the deflections arising from each dP. An expression for the total deflection of the surface $u_z = f(r,\theta)$ is obtained by expressing the local coordinates s and ϕ in terms of r and θ. Thus, substituting Eq. 5.3.3a into 5.3.2d gives u_z in terms of r and θ and where S is the area of the surface of contact:

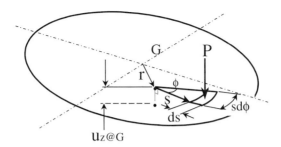

Fig. 5.3.3 Deflection of a general point on the surface is found from the sum of the deflections due to distributed point loads P, the sum of which characterizes the contact pressure distribution.

$$u_z = \frac{1-\nu^2}{\pi E} \iint_S p(s,\phi)\,ds\,d\phi \qquad (5.3.3b)$$

or:

$$u_z = \frac{1-\nu^2}{\pi E} \iint_S p(r,\theta)\,dr\,d\theta \qquad (5.3.3c)$$

Note that the local coordinates s and ϕ correspond to the coordinates r,θ when the point G under consideration is at the axis of symmetry $r = 0$. The strains may be computed from Eq. 5.2.3c and the stresses from Hooke's law, Eq. 5.2.3b. Using this procedure, it can be shown for example that a pressure distribution of the form given by Eq. 5.2a gives rise to displacements beneath the contact circle corresponding to that of a spherical indenter. For a given contact pressure distribution, equations for the stresses within the interior of the specimen may be formulated from a superposition of the Boussinesq field given by Eq. 5.3.2a. Alternatively, one may prescribe the displacements of the surface beneath the circle of contact and, for axis-symmetric indenters, employ integral transform methods[7] to determine the stresses.

5.4 Indentation Stress Fields

We are now in a position to examine indentation stress fields of practical interest. In sections to follow, formulas are presented, without derivation, which give the stresses and deflections of points both on the surface and in the interior of the specimen for a variety of prescribed contact pressure distributions. Our attention will be focused on those indentation configurations for the data shown in Fig. 5.4.1, namely, axis-symmetric spherical, cylindrical punch, and conical indenters. We shall also consider the case of a uniform pressure. Specific attention to the role of the elastic properties of the indenter is discussed in Chapter 6.

In all of the formulas to be presented, the coordinates r and z are positive quantities with positive z corresponding to the direction from the surface into the bulk of the solid. A positive value of displacement u_z indicates a displacement into the bulk of the specimen. A positive value for u_r indicates a displacement away from the axis of symmetry. The contact pressure p_m and radius of circle of contact a are positive quantities. For the stresses, a negative value indicates compression and a positive value indicates tension.

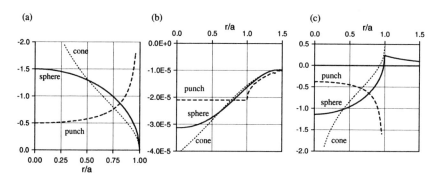

Fig. 5.4.1 (a) Normalized contact pressure distribution σ_z/p_m for spherical indenter, cylindrical punch, and conical indenters. (b) Deflection of the surface spherical, cylindrical, and conical indenters. Deflections in mm calculated for p_m = 1 MPa and radius of circle of contact = 1 mm and for E = 70000 MPa. (c) Magnitude of normalized surface radial stress σ_r/p_m for spherical, cylindrical, and conical indenters. Calculated for $v = 0.26$.

5.4.1 Uniform pressure

In this case, the contact pressure distribution is simply $p = p_m$. For the case of uniform pressure acting over a circular area of radius a, we have:

$$\sigma_z = -p_m \quad r \le a. \tag{5.4.1a}$$

The displacement of the surface, at a general point (r,θ) within the contact circle, is given by Eq. 5.3.3c with $p(r,\theta) = p_m$ in which case we have:

$$u_z = \frac{4(1-v^2)}{\pi E} p_m a \int_0^{\pi/2} \sqrt{1 - \frac{r^2}{a^2}\sin^2\theta}\, d\theta \tag{5.4.1b}$$

where the integral is a complete elliptical integral of the second kind, solutions for which are cumbersome except for the simplest of cases. For example, beneath the indenter at the center of contact r = 0, the normal displacement measured with respect to the original specimen surface is given by[9,12]:

$$u_z = \frac{2(1-v^2)}{E} p_m a \quad r = 0. \tag{5.4.1c}$$

At the edge of the contact circle, r = a, we have[9,12]:

$$u_z = \frac{4(1-v^2)}{\pi E} p_m a \quad r = a. \tag{5.4.1d}$$

Outside the contact circle, the normal displacement can only be calculated using elliptical integrals, solutions of which are beyond the scope of this book but have the form:

$$u_z = \frac{4(1-v^2)}{\pi E} p_m \int_0^{\pi/2} \frac{a^2 \cos^2 \lambda}{r\left(1 - \frac{a^2}{r^2}\sin^2 \lambda\right)} d\lambda \quad r>a. \tag{5.4.1e}$$

The radial and hoop stresses on the surface are[9]:

$$\frac{\sigma_r}{p_m} = \frac{\sigma_\theta}{p_m} = -\frac{(1+2v)}{2} \quad r<a. \tag{5.4.1f}$$

Outside the contact circle, use of elliptical integrals is again required. These stresses are determined from Eq. 5.4.1e and derivatives in Eq. 5.3.2c.

The radial displacements at the surface are given by[9,14]:

$$u_r = -\frac{(1-2v)(1+v)}{2E} p_m r \quad r \leq a \tag{5.4.1g}$$

and[9]:

$$u_r = -\frac{(1-2\nu)(1+\nu)}{2E} p_m \frac{a^2}{r} \quad r > a \tag{5.4.1h}$$

where it will be noticed that u_r for $r > a$ in the present case is the same as that for spherical, cylindrical, and conical indenters given later by Eqs. 5.4.2g, 5.4.4f, and 5.4.5j.

For points within the specimen, along the z axis at $r = 0$, Timoshenko and Goodier[12] give:

$$\frac{\sigma_z}{p_m} = -1 + \frac{z^3}{a^3}\left(1+\frac{z^2}{a^2}\right)^{-3/2} \quad r = 0. \tag{5.4.1i}$$

$$\frac{\sigma_r}{p_m} = \frac{\sigma_\theta}{p_m} = \frac{1}{2}\left[-(1+2\nu)+2(1+\nu)\frac{z}{a}\left(1+\frac{z^2}{a^2}\right)^{-1/2} - \frac{z^3}{a^3}\left(1+\frac{z^2}{a^2}\right)^{-3/2}\right] \tag{5.4.1j}$$

For points within the specimen, Sneddon[15] gives:

$$\frac{\sigma_z}{p_m} = -\frac{1}{2}\left[\frac{\left(3\frac{z}{a}+1\right)\left(\frac{z}{a}+1\right)^2 + \frac{r^2}{a^2}}{\left[\frac{r^2}{a^2}+\left(1+\frac{z}{a}\right)^2\right]^{5/2}}\right] \tag{5.4.1k}$$

An analytical expression for the stress distribution at general points within the specimen half-space is difficult to obtain since the integral in Eq. 5.4.1b is an elliptical integral which cannot be solved for all but the most convenient coordinates, e.g., $z = 0$ or $r = 0$. The complexities of an analytical result may be conveniently bypassed by use of the finite-element method, and the stress distribution within the interior as calculated using this procedure is shown in Fig. 5.4.2.

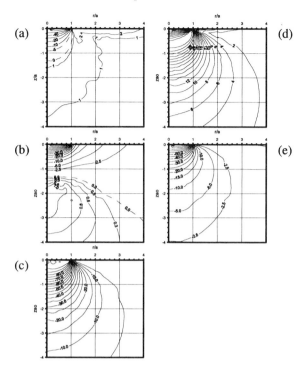

Fig. 5.4.2 Contours of equal stress for uniform pressure calculated for Poisson's ratio ν = 0.26. Distances r and z normalized to the contact radius a and stresses expressed in terms of 100 times the mean contact pressure p_m. (a) σ_1, (b) σ_2, (c) σ_3, (d) τ_{max}, (e) σ_H.

5.4.2 Spherical indenter

The normal pressure distribution directly beneath a spherical indenter was given by Hertz:

$$\frac{\sigma_z}{p_m} = -\frac{3}{2}\left(1-\frac{r^2}{a^2}\right)^{1/2} \quad r \leq a. \tag{5.4.2a}$$

As shown in Fig. 5.4.1, for the pressure distribution given by Eq. 5.4.2a, $\sigma_z = 1.5 p_m$ is a maximum at the center of contact and is zero at the edge of the contact circle. Outside the contact circle, the normal stress σ_z is zero, it being a free surface.

The displacement of points on the surface of the specimen within the contact circle, measured with respect to the original specimen free surface, is[9]:

$$u_z = \frac{1-v^2}{E}\frac{3}{2}p_m\frac{\pi}{4a}\left(2a^2 - r^2\right) \quad r \leq a \tag{5.4.2b}$$

and outside the contact circle is*:

$$u_z = \frac{1-v^2}{E}\frac{3}{2}p_m\frac{1}{2a}\left[\left(2a^2 - r^2\right)\sin^{-1}\frac{a}{r} + r^2\frac{a}{r}\left(1-\frac{a^2}{r^2}\right)^{1/2}\right] \quad r \geq a \tag{5.4.2c}$$

Equations 5.4.2b and 5.4.2c give displacements of points on the surface of the specimen subjected to the pressure distribution given by Eq. 5.4.2a. For the special case of a perfectly rigid indenter, Eq. 5.4.2b evaluated at r = 0 gives the penetration depth beneath the original specimen surface and also the distance of mutual approach between two distance points in both the indenter and specimen. Generally, however, the elastic deformations of the indenter must also be considered (see Chapter 6). Equations 5.4.2b shows that the depth beneath the original surface of the contact circle (at r/a = 1) is exactly one-half of the total depth at r = 0. The displacements of points on the surface calculated using Eqs. 5.4.2b and 5.4.2c are shown in Fig. 5.4.1.

Inside the contact circle, the radial stress distribution at the surface is:

$$\frac{\sigma_r}{p_m} = \frac{1-2v}{2}\frac{a^2}{r^2}\left[1-\left(1-\frac{r^2}{a^2}\right)^{3/2}\right] - \frac{3}{2}\left(1-\frac{r^2}{a^2}\right)^{1/2} \quad r \leq a \tag{5.4.2d}$$

and on the surface outside the contact circle:

$$\frac{\sigma_r}{p_m} = \frac{(1-2v)}{2}\frac{a^2}{r^2} \quad r > a \tag{5.4.2e}$$

It can be shown that the radial displacements, and hence the radial stresses, on the surface outside the contact circle are the same for any symmetrical distribution of pressure within the contact circle; i.e., Eq. 5.4.2e applies also to cylindrical and conical indenters for r > a. The maximum value of σ_r occurs at r = a. The radial stresses on the specimen surface calculated using Eqs. 5.4.2d and 5.4.2e are shown in Fig. 5.4.1. Displacements of points on the surface beneath the indenter in the radial direction are given by[9]:

$$u_r = -\frac{(1-2v)(1+v)a^2}{3E}\frac{3}{r}\frac{3}{2}p_m\left[1-\left(1-\frac{r^2}{a^2}\right)^{3/2}\right] \quad r \leq a \tag{5.4.2f}$$

---

* Reference 9, Eq. 3.42a incorrectly shows a minus sign.

Chapter 5. Elastic Indentation Stress Fields

Note that for all values of r < a, the displacement of points on the surface is inward toward the center of contact. Outside the contact area, the radial displacements are the same as those given previously (Eq. 5.4.1i) for the case of uniform pressure and are given by:

$$u_r = -\frac{(1-2\nu)(1+\nu)a^2}{3E}\frac{3}{r}\frac{p_m}{2} \quad r>a \tag{5.4.2g}$$

The hoop stress, on the surface, is always a principal stress and outside the contact circle is equal in magnitude to the radial stress:

$$\sigma_\theta = -\sigma_r \quad r>a \tag{5.4.2h}$$

Within the interior of the specimen, the stresses are calculated from[3,16]:

$$\frac{\sigma_r}{p_m} = \frac{3}{2}\left\{\frac{1-2\nu}{3}\frac{a^2}{r^2}\left[1-\left(\frac{z}{u^{1/2}}\right)^3\right] + \left(\frac{z}{u^{1/2}}\right)^3\frac{a^2 u}{u^2+a^2 z^2} + \frac{z}{u^{1/2}}\left[u\frac{1-\nu}{a^2+u}+(1+\nu)\frac{u^{1/2}}{a}\tan^{-1}\left(\frac{a}{u^{1/2}}\right)-2\right]\right\} \tag{5.4.2i}$$

$$\frac{\sigma_\theta}{p_m} = -\frac{3}{2}\left\{\frac{1-2\nu}{3}\frac{a^2}{r^2}\left[1-\left(\frac{z}{u^{1/2}}\right)^3\right] + \frac{z}{u^{1/2}}\left[2\nu+u\frac{1-\nu}{a^2+u}-(1+\nu)\frac{u^{1/2}}{a}\tan^{-1}\left(\frac{a}{u^{1/2}}\right)\right]\right\} \tag{5.4.2j}$$

$$\frac{\sigma_z}{p_m} = -\frac{3}{2}\left(\frac{z}{u^{1/2}}\right)^3\left(\frac{a^2 u}{u^2+a^2 z^2}\right) \tag{5.4.2k}$$

$$\frac{\tau_{rz}}{p_m} = -\frac{3}{2}\left(\frac{rz^2}{u^2+a^2 z^2}\right)\left(\frac{a^2 u^{1/2}}{a^2+u}\right) \tag{5.4.2l}$$

where:

$$u = \frac{1}{2}\left[\left(r^2+z^2-a^2\right)+\left[\left(r^2+z^2-a^2\right)^2+4a^2 z^2\right]^{1/2}\right] \tag{5.4.2m}$$

It should be noted that for z = 0, and for values of r/a < 1, the value for u given by Eq. 5.4.2m is always zero but the state of stress directly beneath the indenter may be calculated with reasonable accuracy by taking a sufficiently small value of z.

The principal stresses in the rz plane are given by:

$$\sigma_{1,3} = \frac{\sigma_r + \sigma_z}{2} \pm \sqrt{\left(\frac{\sigma_r - \sigma_z}{2}\right)^2 + \sigma_{rz}^2}$$

$$\sigma_2 = \sigma_\theta$$

$$\tau_{max} = \frac{1}{2}[\sigma_1 - \sigma_3]$$

(5.4.2n)

The angle θ_p between the normal of the plane over which σ_1 is acting and the r axis (surface of the specimen—see Fig. 1.1.10) is found from:

$$\tan \theta_p = -\frac{\sigma_r - \sigma_z}{2\tau_{rz}} \pm \left[\left(\frac{\sigma_r - \sigma_z}{2\tau_{rz}}\right)^2 + 1\right]^{1/2}$$

(5.4.2o)

where ± is the sign of τ_{rz}. In all the formulas involving angles in this book, a regular x-y (or r-z) coordinate system is assumed in which the top right quadrant is +ve x (or r) and +ve y (or z) even though it is customary to show indentation stress fields using the bottom right quadrant (with −ve z). Thus, in Eq. 5.4.2o, a positive value of θ_p is taken from the +ve r axis in an anticlockwise direction to the line of action of σ_1 (see Fig. 1.1.10).

On the surface (at z = 0 and all values of r/a), and also beneath the indenter along the z axis at r = 0, σ_r, σ_θ, and σ_z are principal stresses. The hoop stress, σ_θ, is always a principal stress because of symmetry. On the surface of the specimen, beneath the indenter (r < a), all three principal stresses are compressive and all have approximately the same magnitude. Outside the contact circle but still on the surface, the first principal stress $\sigma_1 = \sigma_r$ is tensile with a maximum value at the edge of the contact circle. This stress is responsible for the formation of Hertzian cone cracks. The second principal stress, $\sigma_2 = \sigma_\theta$, is a hoop stress and is compressive in this region. Outside the contact area along the surface, $\sigma_2 = -\sigma_1$, and beneath the surface along the axis of symmetry at r = 0, the two principal stresses are equal, $\sigma_2 = \sigma_1$. The magnitude of σ_3 at the surface is zero outside the contact circle since it acts normal to a free surface in this region. Along the surface, at all values of r/a, $\sigma_1 = \sigma_r$ and acts in a radial direction. σ_2 is of course a hoop stress, and σ_3 acts normal to the surface.

It is convenient to label the stresses such that $\sigma_1 > \sigma_2 > \sigma_3$ nearly always[†]. Fig. 5.4.3 shows contours of equal values of stress calculated using Eqs. 5.4.2i to 5.4.2n. Note that the contours shown in Figs. 5.4.3 (a) to (e) give no information

---

[†] In general, principal stresses are defined so that $\sigma_1 > \sigma_2 > \sigma_3$. In the indentation stress field, $\sigma_3 > \sigma_2$ at some points just below the surface, but it is usual to specify σ_2 as the hoop stress throughout. Thus, $\sigma_1 > \sigma_2 > \sigma_3$ nearly always.

about the direction or line of action of these stresses. Such information is only available by examining stress trajectories. Stress trajectories are curves whose tangents show the direction of one of the principal stresses at the point of tangency and are particularly useful in visualizing the directions in which the principal stresses act. The stress trajectories of σ_2, being a hoop stress, are circles around the z axis. Stress trajectories for σ_1 and σ_3 can be determined from Eq. 5.4.2o and are shown in Figs. 5.4.3 (f) and (g).

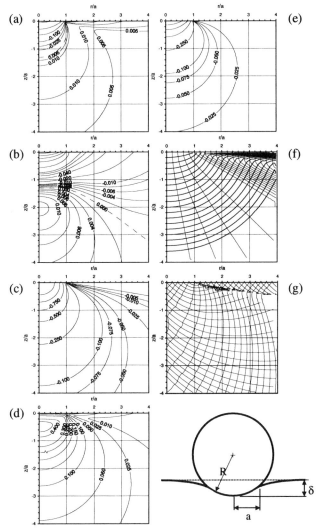

Fig. 5.4.3 Stress trajectories and contours of equal stress for spherical indenter calculated for Poisson's ratio $v = 0.26$. Distances r and z normalized to the contact radius a and stresses expressed in terms of the mean contact pressure p_m. (a) σ_1, (b) σ_2, (c) σ_3, (d) τ_{max}, (e) σ_H, (f) σ_1 and σ_3 trajectories, (g) τ_{max} trajectories.

The feature of the indentation stress field that is important for the initiation of a conical fracture in brittle materials is the tensile region in the specimen surface just outside the area of contact. When the load on the indenter is sufficient, the characteristic cone crack that forms appears to start close to the circle of contact where σ_1 is greatest and proceeds down and outward. Cracks in brittle solids tend to follow a direction of orthogonality with the greatest value of tensile stress; i.e., σ_1. Thus, it is not surprising to observe that cone cracks, as they travel downward into the specimen, appear to follow the σ_3 stress trajectory, which is orthogonal to the σ_1 trajectory. However, in many materials, there is a disparity between the path delineated by this calculated stress trajectory and that taken by the conical portion of an actual crack. Calculations show that, in glass with Poisson's ratio = 0.21, the angle of the cone crack, if it were to follow a direction given by the σ_3 trajectory, should make an angle of approximately 33° to the specimen surface. The actual angle is dependent on Poisson's ratio. However, experimental evidence is that the angle is much shallower, by up to 10° less in some cases. Lawn, Wilshaw, and Hartley[16] attempted to resolve this disparity by analytical computation but were unsuccessful. Yoffe[17] predicts that the answer lies in the modification to the pre-existing stress field by the presence of the actual cone crack as it progresses through the solid, and Lawn[18] proposes a change in local elastic properties in the vicinity of the highly stressed crack tip. As can be seen in Fig. 5.4.3, the position of maximum shear occurs in the specimen at a depth $\approx 0.5a$ and has a maximum value of about $0.49 p_m$ (for $\nu = 0.22$). Plastic deformation in contact loading usually occurs in ductile materials as a result of these shear stresses.

5.4.3 Cylindrical roller (2-D) contact

In the previous section, we summarized the elastic equations for the three-dimensional indentation of a flat-plane, semi-infinite half-space subjected to a pressure distribution associated with a spherical indenter. The two-dimensional analogue of this is the line loading associated with an infinitely long cylindrical "roller." In this case, the indenter load P is given in units of force per unit thickness of the specimen. Johnson[9] shows that the pressure distribution over the area of contact is:

$$\frac{\sigma_z}{p_m} = -\frac{4}{\pi}\left(1 - \frac{x^2}{a^2}\right)^{1/2} \qquad (5.4.3a)$$

where x is the horizontal distance from the axis of symmetry (equivalent to r in axis-symmetric three-dimensional cases). Along the z, or vertical, axis in the interior of the specimen:

$$\frac{\sigma_x}{p_m} = -\frac{4}{\pi}\left[\left(1+2\frac{z^2}{a^2}\right)\left(1+\frac{z^2}{a^2}\right)^{-1/2} - 2\frac{z}{a}\right] \quad (5.4.3b)$$

$$\frac{\sigma_z}{p_m} = -\frac{4}{\pi}\left(1+\frac{z^2}{a^2}\right)^{-1/2} \quad (5.4.3c)$$

σ_x and σ_z given above are principal stresses on the axis of symmetry x = 0.

5.4.4 Cylindrical (flat punch) indenter

The stress field generated by indentation of a flat surface by a cylindrical flat punch is similar to that involving the classical Hertzian stress field. In many ways, a flat punch geometry is preferred over that of a sphere since the contact radius is a constant, independent of the indenter load, thus reducing the number of variables to be analyzed. Further, with a cylindrical punch indenter, the onset of multiple cone cracking, which occurs with a spherical indenter under the expanding contact area, is avoided. However, the sharp edge of a cylindrical punch indenter leads to a singularity in the stress field at the edge of the circle of contact. This leads to plastic deformation of either the specimen or the indenter material. However, if the load on the indenter is not too large, then a small amount of plastic deformation at the edge of the contact circle does not appreciably affect the elastic stress distribution within the specimen material.

The stress field associated with a cylindrical punch indenter has been determined analytically by Sneddon and others[14,19], with a more recent treatment by Barquins and Maugis[20]. The stress field is computed by a superposition of the Boussinesq stress field according to the distribution of pressure beneath the indenter. In the case of a rigid cylindrical flat punch, the contact pressure distribution is:

$$\frac{\sigma_z}{p_m} = -\frac{1}{2}\left(1-\frac{r^2}{a^2}\right)^{-1/2} \quad \text{for } r \le a. \quad (5.4.4a)$$

As shown in Fig. 5.4.1 (a), $\sigma_z = 0.5 p_m$ is a minimum at the center of contact and approaches infinity at the edge. Outside the indenter, $\sigma_z = 0$ along the surface. Beneath the indenter, u_z is the penetration depth beneath the original specimen free surface and is found from[9]:

$$u_z = \frac{1-\nu^2}{E} p_m a \frac{\pi}{2} \quad r \le a \quad (5.4.4b)$$

which is independent of r. Outside the contact circle‡, the normal displacement is [9,14,20]:

$$u_z = \frac{(1-v^2)}{E} p_m a \sin^{-1}\frac{a}{r} \quad r>a \quad (5.4.4c)$$

The displacements for points on the surface calculated using Eqs. 5.4.4b and 5.4.4c are shown in Fig. 5.4.1 (b).

For a cylindrical punch indenter, the radial stress on the surface is:

$$\frac{\sigma_r}{p_m} = \frac{(1-2v)}{2}\frac{a^2}{r^2}\left[1-\left(1-\frac{r^2}{a^2}\right)^{1/2}\right] - \frac{1}{2}\left(1-\frac{r^2}{a^2}\right)^{-1/2} \quad r \leq a \quad (5.4.4d)$$

Outside the contact circle, the radial tensile stress along the surface for a cylindrical punch indenter is precisely the same as that for a spherical indenter and is given by Eq. 5.4.2e, and the radial stresses so computed are shown in Fig. 5.4.1 (c). The radial displacements on the surface are given by[14]:

$$u_r = -\frac{(1-2v)(1+v)a^2}{3E}\frac{3}{r 2}p_m\left[1-\left(1-\frac{r^2}{a^2}\right)^{1/2}\right] \quad r \leq a \quad (5.4.4e)$$

and

$$u_r = -\frac{(1-2v)(1+v)a^2}{3E}\frac{3}{r 2}p_m \quad r > a \quad (5.4.4f)$$

Note that Eqs. 5.4.1i, 5.4.2g, and 5.4.4f are identical. The radial stresses and displacements on the surface outside the circle of contact for both spherical and cylindrical punch indenters are equivalent for the same value of p_m.[21] Note also that the direction of positive z is in the direction of the application of load.

The stress distribution within the specimen in cylindrical coordinates, in terms of the mean contact pressure $p_m = P/\pi a^2$ and the contact radius a, is:

$$\frac{\sigma_r}{p_m} = -\frac{1}{2}\left[J_1^0 - \frac{z}{a}J_2^0 - (1-2v)\frac{a}{r}J_0^1 + \frac{z}{r}J_1^1\right] \quad (5.4.4g)$$

$$\frac{\sigma_\theta}{p_m} = -\frac{1}{2}\left[2vJ_1^0 + (1-2v)\frac{a}{r}J_0^1 - \frac{z}{r}J_1^1\right] \quad (5.4.4h)$$

$$\frac{\sigma_z}{p_m} = -\frac{1}{2}\left[J_1^0 + \frac{z}{a}J_2^0\right] \quad (5.4.4i)$$

‡ Equation 3.38 in reference 9 incorrectly contains the ratio r/a.

$$\frac{\tau_{rz}}{p_m} = -\frac{1}{2}\frac{z}{a}J_2^1 \qquad (5.4.4j)$$

where:

$$J_1^0 = R^{-1/2}\sin\frac{\phi}{2}$$

$$J_0^1 = \frac{a}{r}\left(1 - R^{1/2}\sin\frac{\phi}{2}\right)$$

$$J_2^1 = \frac{r}{a}R^{-3/2}\sin\frac{3\phi}{2}$$

$$J_2^0 = \left[1+\frac{z^2}{a^2}\right]^{1/2} R^{-3/2}\sin\left(\frac{3\phi}{2} - \theta\right)$$

$$J_1^1 = \left[1+\frac{z^2}{a^2}\right]^{1/2} \frac{a}{r} R^{-1/2}\sin\left(\theta - \frac{\phi}{2}\right)$$

$$R = \left[\left[\frac{r^2}{a^2} + \frac{z^2}{a^2} - 1\right]^2 + 4\frac{z^2}{a^2}\right]^{1/2}$$

$$\tan\phi = 2\frac{z}{a}\left[\frac{r^2}{a^2} + \frac{z^2}{a^2} - 1\right]^{-1} \ ; \ \tan\theta = \frac{a}{z} \qquad (5.4.4k)$$

For $\tan\phi > 0$, $0 < \phi < \pi/2$, and for $\tan\phi < 0$, $\pi/2 < \phi < \pi$[20]. Note that the paper by Sneddon[14] contains misprints corrected by Barquins and Maugis[20]. Note also that the factor R in Eqs. 5.4.4k is equivalent to u in Eq. 5.4.2m. Principal stresses and maximum shear stresses can be found by substituting Eqs. 5.4.4i to 5.4.4l into Eqs. 5.4.2n where appropriate. The hoop stress on the specimen surface, is always a principal stress and outside the contact circle is given by Eq. 5.4.2h.

Figures 5.4.4 (a) to (e) shows contours of equal stress, normalized to the mean contact pressure, for a cylindrical punch indenter. Figures 5.4.4 (f) and (g) show the stress trajectories for σ_1, σ_3 and τ_{max}. A comparison between Fig. 5.4.3 and Fig. 5.4.4 shows that the indentation stress fields for a spherical indenter and a cylindrical punch indenter are very similar—except for near the edge of the contact circle. A stress singularity exists at the edge of the contact circle for the cylindrical punch indenter which, in practice, is avoided by localized plastic deformation of either the indenter or the specimen.

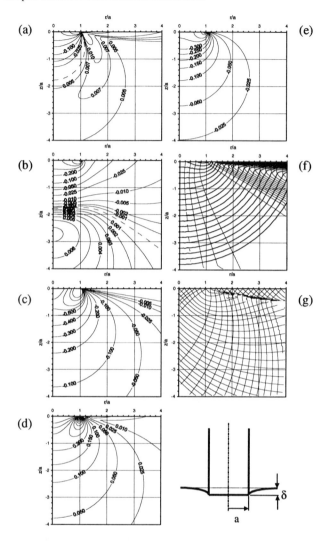

Fig. 5.4.4 Stress trajectories and contours of equal stress for cylindrical flat punch indenter calculated for Poisson's ratio $\nu = 0.26$. Distances r and z normalized to the contact radius a and stresses expressed in terms of the mean contact pressure p_m. (a) σ_1, (b) σ_2, (c) σ_3, (d) τ_{max}, (e) σ_H, (f) σ_1 and σ_3 trajectories, (g) τ_{max} trajectories.

Care should be taken in the use of terminology with this type of indenter because the term "cylindrical indenter" may be taken to mean line contact, such as in a cylindrical roller bearing. The term "cylindrical punch indenter" is preferred and removes any ambiguity.

5.4.5 Rigid cone

The stresses within a semi-infinite half-space loaded by a rigid conical indenter are of significant practical interest because this approximates that used in various hardness tests to be described in Chapter 9. The solutions presented here[22] are similar in format to those presented previously for the case of a cylindrical flat punch indenter. The contact pressure distribution is shown in Fig. 5.4.1 (a) and is given by:

$$\frac{\sigma_z}{p_m} = -\cosh^{-1}\frac{a}{r} \quad r \le a \tag{5.4.5a}$$

In all the cases to be presented in this section, α is the cone semi-angle and quantity $a \cot \alpha$, the depth of penetration measured with respect to the radius of the contact circle (see Fig. 10.4.2). In equations to follow, $\cot \alpha$ may be expressed in terms of the mean contact pressure as:

$$\cot \alpha = \frac{p_m 2(1-v^2)}{E} \tag{5.4.5b}$$

Note that the mean contact pressure depends only on the cone angle and is independent of the load P.

Beneath the indenter, the displacement beneath the original specimen surface is given by[22]:

$$u_z = \left(\frac{\pi}{2} - \frac{r}{a}\right) a \cot \alpha \quad r \le a \tag{5.4.5c}$$

The depth of the circle of contact beneath the specimen surface is given by Eq. 5.4.5c with $r/a = 1$. Outside the contact circle $r > a$, the normal displacement is:

$$u_z = a \cot \alpha \left[\sin^{-1}\frac{a}{r} + \left(\frac{r^2}{a^2} - 1\right)^{1/2} - \frac{r}{a} \right] \quad r > a \tag{5.4.5d}$$

Displacements of points on the surface calculated using Eqs. 5.4.5c and 5.4.5d are shown in Fig. 5.4.1 (b).

The stresses on the surface are:

$$\frac{\sigma_\theta}{p_m} = 2v\frac{\sigma_z}{p_m} - (1-2v)\frac{a}{r}J_0^1 \tag{5.4.5e}$$

and

$$\sigma_r = (1+2v)\sigma_z - \sigma_\theta \tag{5.4.5f}$$

where σ_z for $r < a$ is given by Eq. 5.15a and $\sigma_z = 0$ for $r \geq a$, and:

$$J_0^1 = \frac{a}{2r}\left[1-\left(1-r^2/a^2\right)^{1/2} + \frac{r^2}{a^2}\ln\frac{1+\left(1+r^2/a^2\right)^{1/2}}{r/a}\right] \quad r < a \quad (5.4.5g)$$

and

$$J_0^1 = \frac{a}{2r} \quad r \geq a \quad (5.4.5h)$$

The stresses σ_r, σ_θ, and σ_z at all points on the surface are principal stresses. Outside the contact circle, on the surface of the specimen, the radial and hoop stresses given by Eqs. 5.4.5e and 5.4.5f are precisely the same as those for a cylindrical and spherical indenter. The radial stress on the surface calculated using Eq. 5.4.5f is shown in Fig. 5.4.1.

The radial displacements for $r < a$ are given by:

$$u_r = \frac{1-2\nu}{4(1-\nu)}a\frac{r}{a}\cot\alpha\left[\ln\left(\frac{r/a}{1+\left(1-r^2/a^2\right)^{1/2}}\right) - \frac{1-\left(1-r^2/a^2\right)^{1/2}}{r^2/a^2}\right] \quad (5.4.5i)$$

and for $r \geq a$:

$$u_r = -\frac{1-2\nu}{4(1-\nu)}a\frac{a}{r}\cot\alpha \quad (5.4.5j)$$

which, by making use of Eq. 5.4.5b, is the same as that for spherical and cylindrical indenters and also the case of uniform pressure.

For points within the interior of the specimen:

$$\frac{\sigma_z}{p_m} = -\left[J_1^0 + \frac{z}{a}J_2^0\right]$$

$$\frac{\sigma_\theta}{p_m} = -\left[2\nu J_1^0 + \frac{a}{r}\left((1-2\nu)J_0^1 - \frac{z}{a}J_1^1\right)\right]$$

$$\frac{\sigma_r}{p_m} = -\frac{2(1-\nu^2)}{1-\nu}J_1^0 - \frac{\sigma_z}{p_m} - \frac{\sigma_\theta}{p_m} \quad (5.4.5k)$$

$$\frac{\tau_{rz}}{p_m} = -\frac{z}{a}J_2^1$$

where:

$$J_2^0 = \left(\frac{r^2}{a^2} + \frac{z^2}{a^2}\right)^{-1/2} - \frac{\cos\phi}{R}$$

$$J_1^1 = \frac{a}{r}\left[\left(\frac{r^2}{a^2} + \frac{z^2}{a^2}\right)^{1/2} - R\cos\phi\right]$$

$$J_2^1 = \frac{a}{r}\left[\frac{(1+z^2/a^2)^{1/2}}{R}\cos(\theta-\phi) - \frac{z}{a}\left(\frac{r^2}{a^2} + \frac{z^2}{a^2}\right)^{-1/2}\right]$$

$$J_1^0 = \frac{1}{2}\ln\left[\frac{R^2 + 2R(1+z^2/a^2)^{1/2}\cos(\theta-\phi) + 1 + z^2/a^2}{\left(z/a + (r^2/a^2 + z^2/a^2)^{1/2}\right)^2}\right]$$

$$J_0^1 = \frac{1}{2}\left[\frac{r}{a}J_1^0 + \frac{a}{r}(1 - R\sin\phi) - \frac{z}{a}J_1^1\right]$$

$$R = \left[\left(\frac{r^2}{a^2} + \frac{z^2}{a^2} - 1\right)^2 + 4\frac{z^2}{a^2}\right]^{1/4}$$

$$\tan\theta = \frac{a}{z} \tag{5.4.51}$$

$$\tan 2\phi = 2\frac{z}{a}\left(\frac{r^2}{a^2} + \frac{z^2}{a^2} - 1\right)^{-1}$$

Contours of equal stress and stress trajectories for a conical indenter are shown in Fig. 5.4.5. As a final note, the expressions for tan 2ϕ and R are correctly stated in reference 22 but are not so in reference 7.

100   Chapter 5. Elastic Indentation Stress Fields

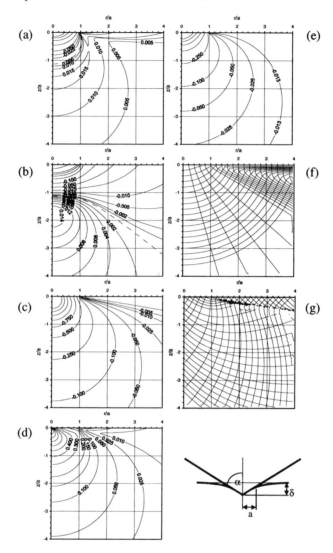

Fig. 5.4.5 Stress trajectories and contours of equal stress for conical indenter calculated for Poisson's ratio $v = 0.26$. Distances r and z normalized to the contact radius a and stresses expressed in terms of the mean contact pressure p_m. (a) σ_1, (b) σ_2, (c) σ_3, (d) τ_{max}, (e) σ_H, (f) σ_1 and σ_3 trajectories, (g) τ_{max} trajectories.

References

1. H. Hertz, "On the contact of elastic solids," J. Reine Angew. Math. 92, 1881, pp. 156–171. Translated and reprinted in English in *Hertz's Miscellaneous Papers*, Macmillan & Co., London, 1896, Ch. 5.
2. H. Hertz, "On hardness," Verh. Ver. Beförderung Gewerbe Fleisses 61, 1882, p. 410. Translated and reprinted in English in *Hertz's Miscellaneous Papers*, Macmillan & Co, London, 1896, Ch. 6.
3. M.T. Huber, "Zur Theorie der Berührung fester elastischer Körper," Ann. Phys. 43 61, 1904, pp. 153–163.
4. S. Fuchs, "Hauptspannungstrajektorien bei der Berührung einer Kugel mit einer Platte," Phys. Z. 14, 1913, pp. 1282–1285.
5. M.T. Huber and S.Fuchs, Phys. Z. 15, 1914, p. 298.
6. W.B. Moreton and L.J. Close, Philos. Mag. 43, 1922, p. 320.
7. I.N. Sneddon, *Fourier Transforms*, McGraw–Hill, New York, 1951.
8. G.M.L. Gladwell, *Contact Problems in the Classical Theory of Elasticity*, Sijthoff and Noordhoff, Alphen aan den Rijn, The Netherlands, 1980.
9. K.L. Johnson, *Contact Mechanics*, Cambridge University Press, Cambridge, U.K., 1985.
10. J. Boussinesq, *Application des Potentiels a l'Etude de l'Equilibre et du Mouvement des Solides Elastiques*, Gauthier-Villars, Paris 1885. Discussed in S.P. Timoshenko and J.N. Goodier, *Theory of Elasticity*, McGraw-Hill, New York, 1970, pp. 398–402.
11. Flamant, C. R., Paris, 114, 1892, p. 1465.
12. S. Timoshenko and J.N. Goodier, *Theory of Elasticity*, 2nd Ed., McGraw-Hill, New York, 1970.
13. J.C. Jaeger and N.G.W. Cook, *Fundamentals of Rock Mechanics*, Chapman and Hall, 1969, London, p. 274.
14. I.N. Sneddon, "Boussinesq's problem for a flat-ended cylinder," Proc. Cambridge Philos. Soc. 42, 1946, pp. 29–39.
15. I.N. Sneddon, "The elastic stresses produced in a thick plate by the application of pressure to its free surfaces," Proc. Cambridge Philos. Soc. 42, 1946, pp. 260–271.
16. B.R. Lawn, T.R. Wilshaw, and N.E.W. Hartley, "A computer simulation study of hertzian cone crack growth," Int. J. Fract. 10 1, 1974, pp. 1–16.
17. E.H. Yoffe, "Elastic stress fields caused by indenting brittle materials," Philos. Mag. A, 46 4, 1982, p. 617.
18. B.R. Lawn, private communication, 1994.
19. J.W. Harding and I.N. Sneddon, "The elastic stresses produced by the indentation of the plane surface of a semi-infinite elastic solid by a rigid punch," Proc. Cambridge Philos. Soc. 41, 1945, pp. 16–26.
20. M. Barquins and D. Maugis, "Adhesive contact of axisymmetric punches on an elastic half-space: the modified Hertz-Huber's stress tensor for contacting spheres," J. Mec. Theor. Appl. 1 2, 1982, pp. 331–357.
21. S. Way, "Some observations on the theory of contact pressures," J. Appl. Mech. 7, 1940, pp. 147–157.
22. I.N. Sneddon, "Boussinesq's problem for a rigid cone," Proc. Cambridge Philos. Soc. 44, 1948, pp. 492–507.

Chapter 6
Elastic Contact

6.1 Hertz Contact Equations

The stresses and deflections arising from the contact between two elastic solids have practical application in hardness testing, wear and impact damage of engineering ceramics, the design of dental prostheses, gear teeth, and ball and roller bearings. In many cases, the contact between a rigid "indenter" and a flat, extensive "specimen" is of particular interest. The most well-known scenario is the contact between a rigid sphere and a flat surface, where Hertz[1,2] found that the radius of the circle of contact a is related to the indenter load P, the indenter radius R, and the elastic properties of the materials by:

$$a^3 = \frac{4}{3}\frac{kPR}{E} \qquad (6.1a)$$

where k is an elastic mismatch factor given by:

$$k = \frac{9}{16}\left[(1-v^2)+\frac{E}{E'}(1-v'^2)\right] \qquad (6.1b)$$

In Eq. 6.1b, E, v and E', v' are the Young's modulus and Poisson's ratio for the specimen and the indenter, respectively*. Hertz also found that the maximum tensile stress in the specimen occurs at the edge of the contact circle at the surface and is given by:

$$\sigma_{max} = (1-2v)\frac{P}{2\pi a^2} \qquad (6.1c)$$

This stress, acting in a radial direction on the surface outside the indenter, decreases as the inverse square of the distance away from the center of contact and is usually considered responsible for the production of Hertzian cone cracks (see Chapter 7). Combining Eqs. 6.1a and 6.1c, the maximum tensile stress outside the indenter can be expressed in terms of the indenter radius R:

---

* Some authors prefer to define k as: $k = 1+\left[(1-v'^2)E/(1-v^2)E'\right]$ in which case we have: $a^3 = 3/4\left(1-v^2\right)kPR/E$.

$$\sigma_{max} = \left(\frac{(1-2\nu)P}{2\pi}\right)\left(\frac{3E}{4k}\right)^{2/3} P^{1/3} R^{-2/3}. \tag{6.1d}$$

The mean contact pressure, p_m, is given by the indenter load divided by the contact area and is a useful normalizing parameter, which has the additional virtue of having actual physical significance.

$$p_m = \frac{P}{\pi a^2}. \tag{6.1e}$$

It can be shown from Eq. 6.1a that the contact area is proportional to $P^{2/3}$ and therefore p_m is proportional to $P^{1/3}$. Substituting Eq. 6.1e into Eq. 6.1a gives[†]:

$$p_m = \left(\frac{3E}{4\pi k}\right)\frac{a}{R}. \tag{6.1f}$$

We may refer to the mean contact pressure as the "indentation stress" and the quantity a/R as the "indentation strain." We may question this definition of "indentation strain", but it is appropriate since the elastic strains within the specimen scale with this ratio. For example, without some external reference, it is not possible to tell the difference between indentations made with indenters of different radii if the quantity a/R is the same in each. The functional relationship between p_m and a/R foreshadows the existence of a stress–strain response similar in nature to that more commonly obtained from conventional uniaxial tension and compression tests. In both cases, a fully elastic condition yields a linear response. However, due to the localized nature of the stress field, an *indentation stress-strain* relationship yields valuable information about the elastic-plastic properties of the test material that is not generally available from uniaxial tension and compression tests.

6.2 Contact Between Elastic Solids

In sections to follow, Hertz's original analysis is reviewed to gain an understanding of the process of indentation and the nature of the contact between elastic solids. The following assumptions serve to facilitate the analysis:

i. The radii of curvature of the contacting bodies are large compared with the radius of the circle of contact. With this assumption, we may treat

---

[†] or
$$p_m = \left(\frac{4E}{3\pi k(1-\nu^2)}\right)\frac{a}{r}$$
using the alternative definition of k.

each surface as an elastic half-space where equations for the stresses and displacements have been given in Chapter 5.

ii. The dimensions of each body are large compared with the radius of the circle of contact. This allows indentation stresses and strains to be considered independently of those arising from the geometry, method of attachment, and boundaries of each solid.

iii. The contacting bodies are in frictionless contact. That is, only a normal pressure is transmitted between the indenter and the specimen. Table 6.1 summarizes the distribution of surface normal pressures to be considered.

Table 6.1 Equations for surface pressure distributions beneath the indenter for different types of indentations.

Indenter type	Equation for normal pressure distribution r < a
Sphere	$\dfrac{\sigma_z}{p_m} = -\dfrac{3}{2}\left(1 - \dfrac{r^2}{a^2}\right)^{1/2}$
Cylinder 2-D	$\sigma_z = -\dfrac{2P}{\pi a}\left(1 - \dfrac{x^2}{a^2}\right)^{1/2}$
Cylindrical Flat punch	$\dfrac{\sigma_z}{p_m} = -\dfrac{1}{2}\left(1 - \dfrac{r^2}{a^2}\right)^{-1/2}$
Uniform pressure	$\sigma_z = -p_m$
Cone	$\dfrac{\sigma_z}{p_m} = -\cosh^{-1}\dfrac{a}{r}$

6.2.1 Spherical indenter

Consider the contact of a sphere of radius R' with elastic modulus E' and Poisson's ratio v' with the flat surface of a specimen of radius R_s whose elastic constants are E and v. With no load applied, and with the indenter just touching the specimen, the distance h between a point on the periphery of the indenter to the specimen surface as a function of radial distance r is given by:

$$h = \frac{r^2}{2R} \qquad (6.2.1a)$$

where R is the relative curvature of the indenter and the specimen:

$$\frac{1}{R} = \frac{1}{R'} + \frac{1}{R_S} \qquad (6.2.1b)$$

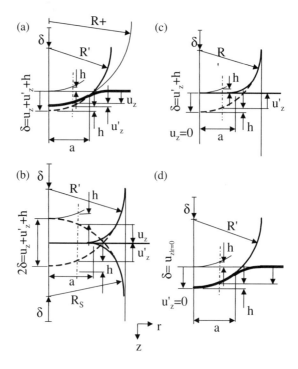

Fig. 6.2.1 Schematic of contact between two elastic solids. (a) Nonrigid spherical indenter and nonrigid, flat specimen; (b) two identical nonrigid spheres; (c) nonrigid spherical indenter and flat, rigid specimen; (d) rigid, spherical indenter and flat, nonrigid specimen (after reference 3).

Now, in Fig. 6.2.1a, load is applied to the indenter in contact with a flat surface (R_s in Eq. 6.2.1b = ∞) such that the point at which load is applied moves a vertical distance δ. This distance is often called the "load-point displacement" and when measured with respect to a distant point in the specimen may be considered the distance of mutual approach between the indenter and the specimen. In general, both the indenter and specimen surface undergo deformation. These deformations are shown in the figure by u'_z and u_z at some arbitrary point inside the contact circle for both the indenter and the specimen respectively. Inspection of Fig. 6.2.1a shows that the load-point displacement is given by:

$$\delta = u'_z + u_z + h \qquad (6.2.1c)$$

If the indenter is perfectly rigid, then $u'_z = 0$ (see Fig. 6.2.1d). For both rigid and nonrigid indenters, h = 0 at r = 0, and thus the load-point displacement is given by $\delta = u'_z + u_z$. Note that u'_z, u_z, and h are all functions of r, although we have yet to specify this function $u_z(r)$ precisely.

Hertz showed that a distribution of pressure of the form given by that for a sphere in Table 6.1 results in displacements of the specimen surface (see Chapter 5) as given by:

$$u_z = \frac{1-v^2}{E} \frac{3}{2} p_m \frac{\pi}{4a}\left(2a^2 - r^2\right) \quad r \leq a \qquad (6.2.1d)$$

After deformation, the contact surface lies between the two original surfaces and is also part of a sphere whose radius depends on the relative radii of curvature of the two opposing surfaces and elastic properties of the two contacting materials. For the special case of contact between a spherical indenter and a flat surface where the two materials have the same elastic properties, the radius of curvature of the contact surface is twice that of the radius of the indenter. The Hertz pressure distribution acts equally on both the surface of the specimen and the indenter, and the deflections of points on the surface of each are thus given by Eq. 6.2.1d‡. Thus, substituting Eq. 6.2.1d into Eq. 6.2.1c for both u'_z and u_z and making use of Eq. 6.2.1a, we obtain, for the general case of a nonrigid indenter and specimen:

$$u'_z + u_z = \left(\frac{1-v'^2}{E'} + \frac{1-v^2}{E}\right)\frac{\pi}{4a}\frac{3}{2}p_m\left(2a^2 - r^2\right)$$

$$= \delta - \frac{r^2}{2R} \qquad (6.2.1e)$$

where R is the relative radius of curvature. With a little rearrangement, and setting r = a in Eq. 6.2.1e, it is easy to obtain the Hertz equation, Eq. 6.1a, and to show that at r = 0, the distance of mutual approach δ between two distant points within the indenter and the specimen is given by:

$$\delta^3 = \left(\frac{4k}{3E}\right)^2 \frac{P^2}{R} \qquad (6.2.1f)$$

where k is as given in Eq. 6.1b. Substituting Eq. 6.1a into 6.2.1f, we have the distance of mutual approach, or load-point displacement, for both rigid and nonrigid indenters as:

$$\delta = \frac{a^2}{R} \qquad (6.2.1g)$$

When the indenter is perfectly rigid, $k = 9(1-v^2)/16$, and the distance of mutual approach δ is equal to the penetration depth $u_{z|r=0}$ below the original speci-

---

‡ The Hertz analysis approximates the curved surface of a sphere as a flat surface since the radius of curvature is assumed to be large in comparison to the area of contact.

men free surface as given by Eq. 6.2.1d. From Eq. 6.2.1d, for both rigid and nonrigid indenters, the depth of the edge of the circle of contact is exactly one half of that of the total depth of penetration beneath the surface; i.e., $u_{z|r=a} = 0.5 u_{z|r=0}$.

Following Johnson[4], we may define the quantity $E^* = 9E/(16k)$ such that:

$$\frac{1}{E^*} = \frac{(1-v^2)}{E} + \frac{(1-v'^2)}{E'} \tag{6.2.1h}$$

where Eq. 6.2.1f can be written:

$$\delta^3 = \left(\frac{3}{4E^*}\right)^2 \frac{P^2}{R}. \tag{6.2.1i}$$

The quantity E^* is the effective elastic modulus of the system and decreases as the indenter becomes less rigid. Thus, for a particular value of load P, the distance of mutual approach δ for a nonrigid indenter is greater than that for a rigid indenter due to the deformation of the indenter. For a spherical indenter, the radius of the circle of contact a also increases with a decreasing value of E^* (or an increasing value of k) as per Eq. 6.1a and hence, for the same value of load P, the mean contact pressure is reduced. Hertz showed that, for contact between two spheres, the profile of the surface of contact was also a sphere with a radius of curvature intermediate between that of the contacting bodies and more closely resembling the body with the greatest elastic modulus. Thus, as shown in Fig. 6.2.1a, contact between a flat surface and a nonrigid indenter of radius R is equivalent to that between the flat surface and a perfectly rigid indenter of a larger radius R^+, which may be computed using Eq. 6.1a with k set as for a rigid indenter. If the contact is viewed in this manner, then the load-point displacement of an equivalent rigid indenter is given by Eq. 6.2.1d with r = 0 and not Eq. 6.2.1i. Thus, in terms of the radius of the contact circle a, the equivalent rigid indenter radius is given by:

$$R^+ = \frac{3a^3 16E}{4(1-v^2)9P}$$

$$= \frac{4Ea}{(1-v^2)3\pi p_m} \tag{6.2.1j}$$

Note that for the special case of the contact between two spheres of equal radii and the same elastic constants, the equivalent rigid indenter radius $R^+ \to \infty$ and the profile of the contact surface is a straight line (see Fig. 6.2.1b).

Finally, it should be noted that for a spherical indenter, the mean contact pressure is proportional to $P^{1/3}$.

6.2.2 Flat punch indenter

For a pressure distribution corresponding to that of a rigid cylindrical indenter, the relationship between load and displacement of the surface u_z relative to the specimen free surface beneath the indenter is:

$$P = \frac{2Ea}{1-v^2} u_z \qquad (6.2.2a)$$

Elasticity of the indenter may be taken into account by replacing E by the smaller quantity E^* (Eq. 6.2.1h). For both rigid and nonrigid indenters, the radius of the circle of contact is a constant and hence so is the mean contact pressure for a given load P (neglecting any localized deformations of the indenter material at its periphery). Therefore, the deflection of points on the specimen surface beneath the indenter must remain unchanged. In this case, in Eq. 6.2.2a, with E = E^*, u_z is the distance of mutual approach between the indenter and the specimen for a given load P. In Eq. 6.2.2a, with E equal to that of the specimen, u_z is the penetration depth. For a rigid indenter, u_z in Eq. 6.2.2a is both the penetration depth and the distance of mutual approach δ as shown in Fig. 6.2.2. Thus, unlike the case of a spherical indenter, the penetration depth for a cylindrical punch indenter is the same for both rigid and nonrigid indenters since the pressure distribution is the same for each case.

Finally, for a cylindrical indenter, the mean contact pressure is directly proportional to the load since the contact radius a is a constant.

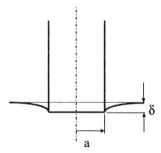

Fig. 6.2.2 Geometry of contact with cylindrical flat punch indenter.

6.2.3 Conical indenter

For a conical indenter, we have[5]:

$$P = \pi a^2 \cot\alpha \frac{E}{2(1-v^2)} \qquad (6.2.3a)$$

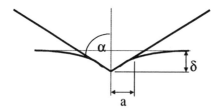

Fig. 6.2.3 Geometry of contact with conical indenter.

and

$$u_z = \left(\frac{\pi}{2} - \frac{r}{a}\right) a \cot \alpha \quad r \leq a \tag{6.2.3b}$$

where α is the cone semiangle as shown in Fig. 6.2.3. Substituting Eq. 6.2.3b with $r = 0$ into 6.2.3a, we obtain:

$$P = \tan \alpha \frac{2E}{\pi(1-v^2)} u_{z|r=0}^2 \tag{6.2.3c}$$

where $u_{z|r=0}$ is the depth of penetration of the apex of the indenter beneath the original specimen free surface. For a nonrigid indenter, replacing E with the smaller quantity E^* means that for the same load P, Eq. 6.2.3c gives the distance of mutual approach between the indenter and the specimen, since the mean contact pressure is a constant and independent of the load (as given in Table 6.1).

6.3 Impact

In many practical applications, the response of brittle materials to projectile impacts is of considerable interest. In most cases, an equivalent static load may be calculated and the indentation stress fields of Chapter 5 applied as required.

The load-point displacement can be expressed in terms of the indenter load P as given in Eq. 6.2.1f. For a sphere impacting on a flat plane specimen, the time rate of change of velocity is related to the mass of the indenter and the load:

$$m \frac{d^2 \delta}{dt^2} = -P \tag{6.3.1}$$

Equating Eqs. 6.2.1f and 6.3.1[6], multiplying both sides by velocity and integrating, we obtain:

$$m\frac{dv}{dt} = -\delta^{3/2} R^{1/2} \frac{3E}{4k}$$

$$mv\frac{dv}{dt} = -\delta^{3/2} R^{1/2} \frac{3E}{4k} \frac{d\delta}{dt}$$

$$\int_{v_o}^{v} mv\,dv = \int -\delta^{3/2} R^{1/2} \frac{3E}{4k} d\delta \qquad (6.3.2)$$

$$\frac{1}{2} mv_o^2 = \frac{2}{5} \delta^{5/2} R^{1/2} \frac{3E}{4k}$$

Equation 6.3.2 equates the kinetic energy of the projectile to the strain potential energy stored in the specimen. Maximum strain energy occurs when the final velocity is zero. Substituting δ from Eq. 6.2.1f, the maximum load at impact is thus:

$$P = \left[\left(\frac{5}{4}\right)^3 \frac{9}{16} \frac{E^2}{k^2} m^3 v_o^6 \right]^{1/5} \qquad (6.3.3)$$

where m is the mass of the indenter, R is the indenter radius, v_o is the indenter velocity, E is the elastic modulus of the specimen, and k is the elastic mismatch constant as given by Eq. 6.1b. The load given by Eq. 6.3.3 can be used for calculating stress fields and displacements for impact loading.

Similar expressions can be developed for cylindrical and conical indenters. For a cylindrical punch indenter, the maximum impact loading is:

$$P = v_o \left[\frac{2Eam}{1-v^2} \right]^{1/2} \qquad (6.3.4)$$

and for a conical indenter, we obtain:

$$P = \left[\tan\alpha \frac{2E}{\pi(1-v^2)} \right]^{1/3} \left[\frac{3}{2} mv_o^2 \right]^{2/3} \qquad (6.3.5)$$

In practice, the nature of cracking of brittle materials under impact loading is dependent on the thickness of the specimen since bending stresses, as well as contact stresses, act on surface flaws. A relationship between impact velocity and specimen thickness is given by Ball[7], who shows that the type of cracking

112   Chapter 6. Elastic Contact

can range from a "star" on the back side of the specimen to completely perfect cones, incomplete cones with crushing, or just small ring cracks.

6.4 Friction

In all the equations presented so far, no account has been made of any effects of friction between the indenter and the specimen surface (i.e., interfacial friction). Indeed, one of the original boundary conditions of Hertz's original analysis was that of frictionless contact. Now, although such an assumption may be acceptable for a large number of cases of practical interest, it is nevertheless important to have some understanding of the effects of interfacial friction for those cases in which friction is an important parameter.

Figure 6.4.1 shows four different scenarios relating to interfacial friction which will facilitate our introductory treatment of this complex phenomenon. Consider two points on the indenter and specimen surfaces which come into contact during an indentation loading. For the purposes of discussion, we shall assume that the indenter and specimen have different elastic properties, with the modulus of the indenter being much larger than that of the specimen. As shown in Fig. 6.4.1 (a), for the condition of full slip (no friction), upon loading, points on both specimen and indenter move inward toward the axis of symmetry under the influence of the applied forces F_a. No friction forces are involved. Movement of points within the specimen material generates "internal" forces F_s (i.e., from the stresses set up in the material) which are proportional to the relative displacement. Movement ceases when the internal forces F_s balance the applied forces. Upon unloading, internal forces diminish as the applied force is decreased. Points on the surface move back to their original positions.

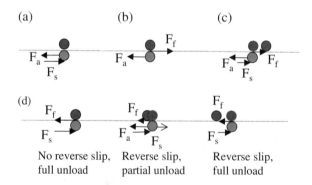

Fig. 6.4.1 Points on the indenter and specimen surfaces that have come into contact during loading. (a) full slip, (b) no slip, (c) partial slip—loading, (d) partial slip—unloading. In (d), reverse slip may occur, leading to residual stresses.

Consider now the case of no slip (i.e., full adhesive contact). As shown in Fig. 6.4.1 (b), upon loading, points on the specimen surface want to move inward under the influence of the applied forces F_a but are prevented from doing so by frictional forces F_f. The applied force F_a is balanced by the friction force F_f. Upon unloading, frictional forces diminish as the applied force is decreased. Points on the two surfaces remain in their original positions.

In the case of partial slip, loading and unloading must be considered separately. Upon loading, Fig. 6.4.1 (c), points on the specimen surface want to move inward under the influence of the applied forces F_a. Some points are prevented from doing so by frictional forces which, due to the local magnitude of the normal forces, are large enough to balance the applied forces. For other points, the applied forces are greater than the frictional force and those points do move inwards—slip occurs between the surfaces. For those points that have slipped, the frictional force has reached its maximum value. Internal forces can still increase with increasing load. Relative movement occurs until the internal force F_s plus the maximum frictional force F_f opposes the applied force F_a. The friction force is now applied by a new point on the indenter, which has now come into contact with the point on the surface of the specimen. Now, at full load, the applied force at a point that has slipped is balanced by the sum of the maximum frictional force and the internal force. The frictional force arises when there is relative shear loading on the contacting surfaces. The magnitude of the frictional force is equal and opposite to the shear force loading and reaches a maximum value dependent on the coefficient of friction and the magnitude of the normal force between the surfaces at that point.

Consider now the forces between two points *that have slipped* during loading, such as shown in Fig. 6.4.1 (d). If the applied force is relaxed slightly, then the frictional force diminishes. No relative movement of the points occurs so the internal forces remain constant. As the load is reduced a little further, the frictional forces reduce and eventually are reduced to zero. At this point, the applied force is balanced entirely by the internal force and there is no shear force between the surfaces. As the load is reduced even further, frictional forces of opposite sign act on the surface. Internal forces are now balanced by both the frictional forces and the applied forces. As the applied load is reduced, the reverse frictional force increases up to a maximum value. No relative movement of the surfaces occur so there is no reduction in the internal forces yet. At the limit of adhesion, the friction force has reached a maximum value and any further reduction in applied load results in relative movement between the surfaces. This has the effect of diminishing the internal forces (internal stresses begin to relax). At this point, reverse slipping is occurring. The friction force remains at its maximum value as the applied load is decreased. As the applied load is reduced to zero, the frictional force remains at its maximum value and is balanced by "residual" internal forces. During unloading, the limiting value of friction force may never be reached (i.e., the applied force is reduced to zero with frictional forces continuing to increase and balance the internal forces). This also results in the specimen containing residual stress (at a larger magnitude than would have resulted if reverse slip had occurred).

The effect of interfacial friction is to create an inner region of full adhesive contact with an outer annulus where the surfaces have slipped. The inner radius of this annulus is called the "slip radius" whereas the outer radius is the radius of the circle of contact. For the case of full slip, the slip radius is zero. For the case of no slip, the slip radius is equal to the contact radius. Analytical treatments of contact with interfacial friction are usually presented for the simplest cases of either full slip, $\mu = 0$, or no slip, $\mu = \infty$[8]. Perhaps the most complete treatment is that of Spence[9], who calculated the distribution of surface stresses for both a sphere and punch. There have also been a number of finite-element studies of frictional contact[10,11,12] reported in the literature.

Consider the case of a spherical indenter for various coefficients of friction. Figure 6.4.2 (a) shows finite-element results undertaken by the author for the variation of radial stress on the specimen surface for a particular indenter radius and loading condition for different coefficients of friction ranging from full slip $\mu = 0$ to no slip or fully bonded contact. For comparison purposes, the radial stresses as computed using Eqs. 5.4.2d and 5.4.2e are shown along with the finite-element results for $\mu = 0$. Note the diminished magnitude of the maximum radial stress just outside the contact radius as the friction coefficient increases. Radial stresses in this region are responsible for the production of Hertzian cone cracks. Due to the reduction in radial stress with increasing values of μ, one may conclude that the probability of a cone crack occurring for a given indenter load may be reduced with an increasing friction coefficient between the indenter and specimen surfaces. Figure 6.4.2 (b) shows radial displacements along the specimen surface. In this figure, the horizontal axis is normalized to the radius of the circle of contact. The contact radius is thus r/a = 1. The slip radius can be readily determined from the point where each line for each value of μ meets the upper horizontal axis. For points on the surface within the slip radius, the radial displacements are very small since the material is constrained from moving inward by frictional forces. Similar behavior is seen with a cylindrical punch indenter, as shown in Fig. 6.4.3.

However, there is one important difference between the relationship between the slip radius and the contact radius for spherical and cylindrical punch indenters. Finite-element results show that for a spherical indenter, the slip radius is dependent on the indenter load, although the ratio of the slip radius to the contact radius remains fairly independent of load but does depend, almost linearly, on the coefficient of friction. For a cylindrical punch indenter, the slip radius is independent of indenter load and only depends (nonlinearly) on the coefficient of friction for the contacting surfaces.

Finally, it should be noted that finite-element results indicate that the stresses σ_z and displacements u_z in the normal direction appear to be unaffected by the presence of interfacial friction.

Chapter 6. Elastic Contact 115

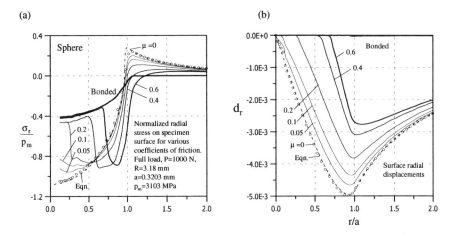

Fig. 6.4.2 Finite-element results for (a) the variation of normalized radial stress and (b) radial displacements (mm) of the specimen surface for a spherical indenter R = 3.18 mm at P = 1000 N. The radius of the circle of contact is a = 0.3203 mm which gives a mean contact pressure p_m = 3103 MPa. Results are shown for different coefficients of friction from full slip μ = 0 to no slip or fully bonded contact.

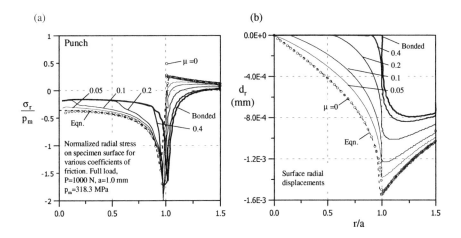

Fig. 6.4.3 Finite-element results for (a) the variation of normalized radial stress and (b) radial displacements (mm) of the specimen surface for a cylindrical punch indenter R = 1 mm at P = 1000 N. The mean contact pressure p_m = 318.3 MPa. Results are shown for different coefficients of friction from full slip μ = 0 to no slip or fully bonded contact.

References

1. H. Hertz, "On the contact of elastic solids," J. Reine Angew. Math. 92, 1881, pp. 156–171. Translated and reprinted in English in *Hertz's Miscellaneous Papers*, Macmillan & Co., London, 1896, Ch. 5.
2. H. Hertz, "On hardness," Verh. Ver. Beförderung Gewerbe Fleisses 61, 1882, p. 410. Translated and reprinted in English in *Hertz's Miscellaneous Papers*, Macmillan & Co, London, 1896, Ch. 6.
3. A.C. Fischer-Cripps, "The hertzian contact surface," J. Mater. Sci. 34, 1999, pp. 129–137.
4. K.L. Johnson, *Contact Mechanics*, Cambridge University Press, Cambridge, U.K., 1985.
5. I.N. Sneddon, "Boussinesq's problem for a rigid cone," Proc. Cambridge Philos. Soc. 44, 1948, pp. 492–507.
6. S. Timoshenko and J.N. Goodier, *Theory of Elasticity*, 2nd Ed., McGraw-Hill, New York, 1970.
7. A. Ball, "On the bifurcation of cone cracks in glass plates," Philos. Mag. A, 73 4, 1996, pp. 1093–1103.
8. K.L. Johnson, J.J. O'Connor and A.C. Woodward, "The effect of indenter elasticity on the Hertzian fracture of brittle materials," Proc. R. Soc. London, Ser. A334, 1973, pp. 95–117.
9. D.A. Spence, "The Hertz contact problem with finite friction," J. Elastoplast. 5 3–4, 1975, pp. 297–319.
10. J. Tseng and M.D. Olson, "The mixed finite element method applied to two-dimensional elastic contact problems," Int. J. Numer. Methods Eng. 17, 1981, pp. 991–1014.
11. N. Okamoto and M. Nakazawa, "Finite element incremental contact analysis with various frictional conditions," Int. J. Numer. Methods Eng. 14, 1979, pp. 337–357.
12. T.D. Sachdeva and C.V. Ramakrishnan, "A finite element solution for the two-dimensional elastic contact problems with friction," Int. J. Numer. Methods Eng. 17, 1981, pp. 1257–1271.

Chapter 7
Hertzian Fracture

7.1 Introduction

The tensile or bending strength of a particular specimen of brittle material may be severely reduced in the presence of cone cracks formed due to contact loading. The conditions required to initiate a cone crack in a brittle material are therefore of significant practical interest. Cone cracks resulting from contact with spherical indenters were first reported in the scientific literature by Hertz in 1881[1,2] and are referred to as Hertzian cone cracks regardless of the indenter type used to produce them. We have examined the nature of the contact stress fields associated with various indenters in Chapter 5 and presented equations for elastic contact in Chapter 6. In this chapter, we investigate those factors that influence the load required to initiate a Hertzian cone crack for a particular indenter size and specimen surface condition.

7.2 Hertzian Contact Equations

As we saw in Chapter 6, Hertz formulated mathematical relationships between indenter load P, indenter radius R, contact area a, and maximum tensile stress, σ_{max}. The contact radius depends on the load, the indenter radius, and the elastic properties of both the specimen and the indenter according to:

$$a^3 = \frac{4}{3}\frac{kPR}{E} \qquad (7.2a)$$

where k is an elastic mismatch factor:

$$k = \frac{9}{16}\left[\left(1-v^2\right)+\frac{E}{E'}\left(1-v'^2\right)\right] \qquad (7.2b)$$

The maximum tensile stress occurs at the edge of the contact circle and is given by:

$$\sigma_{max} = (1-2v)\frac{P}{2\pi a^2} \qquad (7.2c)$$

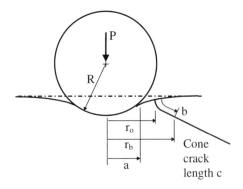

Fig. 7.2.1 Geometry of Hertzian cone crack. The crack begins normal to the specimen surface and extends down a small distance before widening into a fully developed cone (after reference 3).

Substituting Eq. 7.2a into Eq. 7.2c we have:

$$\sigma_{max} = \left(\frac{(1-2\nu)}{2\pi}\right)\left(\frac{3E}{4k}\right)^{2/3} P^{1/3} R^{-2/3} \tag{7.2d}$$

The tensile stress on the specimen surface near the edge of the circle of contact is usually responsible for the production of Hertzian cone cracks. As shown in Fig. 7.2.1, Hertzian cone cracks generally consist of an initial ring (normal to the specimen free surface) which extends a very short distance into the specimen before evolving into a cone, the angle of which depends on Poisson's ratio of the material and also on the method of specimen support and thickness.

7.3 Auerbach's Law

During an experimental investigation into the hardness of materials, Auerbach[4] in 1891 found that for a wide range of brittle materials, the force P required to produce a cone crack was proportional to the radius of the indenter R such that:

$$P = AR \tag{7.3a}$$

where A is termed the Auerbach constant. Equation 7.3a is an empirical result (i.e., based upon experimental observations without any explanation as to its physical cause), which has become known as "Auerbach's law."

Equation 7.3a can be alternatively written in terms of the radius of the contact area a, using Eq. 7.2a:

$$P = \left(\frac{3}{4}\frac{AE}{k}\right)^{1/2} a^{3/2} \tag{7.3b}$$

and substituting Eq. 7.3a into 7.2d gives:

$$\sigma_{max} = \left(\frac{(1-2v)A^{1/3}}{2\pi}\right)\left(\frac{3E}{4k}\right)^{2/3} R^{-1/3}. \tag{7.3c}$$

If σ_{max} is the maximum tensile stress upon the occurrence of a cone crack, then Auerbach's law appears to imply that the tensile strength of material depends on the radius of the indenter rather than being a material property—a size effect worthy of special note.

7.4 Auerbach's Law and the Griffith Energy Balance Criterion

The Griffith[5] criterion for fracture relates the energy needed to form new crack surfaces and the attendant release in strain energy (see Chapter 3). The externally applied uniform stress σ_a required for the growth of an existing flaw of length 2c and unit width in an infinite solid is given by:

$$\sigma_a \geq \left(\frac{2\gamma E}{(1-v^2)\pi c}\right)^{1/2} \tag{7.4a}$$

where γ is the *fracture* surface energy in J m^{-2}. The $1-v^2$ term is included for the general case of plane strain. Equation 7.4a applies directly to a double-ended crack of length 2c contained fully within a uniformly stressed solid where the stress is applied normal to the crack. It may also be applied with only a small error to a half crack of length c, such as a surface flaw.

The Griffith criterion is more commonly stated in terms of Irwin's stress intensity factor[6] K_1, where:

$$\frac{K_1^2(1-v^2)}{E} \geq 2\gamma \tag{7.4b}$$

where, for the case of an infinite solid:

$$K_1 = \sigma\sqrt{\pi c}. \tag{7.4c}$$

The left-hand side of Eq. 7.4b is termed the strain energy release rate and is given the symbol G. The Griffith criterion is satisfied for $K_1 \geq K_{1C}$, where K_{1C} may be considered a material property which can be readily measured in the laboratory. A typical value for soda-lime glass is 0.78 MPa m$^{1/2}$. Using this value, Eqs. 7.4b and 7.4c give a fracture surface energy for soda-lime glass of $\gamma = 3.6$ J m^{-2}, which is in agreement with various experimentally determined values of this quantity[7].

Early workers applied the Griffith fracture criterion to flaws in the vicinity of an indenter in terms of the surface tensile stress only, as given by the Hertz equations. Using this method, a combination of Eqs. 7.2d and 7.4a gives the critical condition for failure as:

$$\left(\frac{2E\gamma}{(1-v^2)\pi c}\right)^{1/2} = \left(\frac{(1-2v)}{2\pi}\right)\left(\frac{3E}{4k}\right)^{2/3}\frac{P^{1/3}}{R^{2/3}} \qquad (7.4d)$$

Equation 7.4d states that P is proportional to $R^2c^{-3/2}$. If all the flaws in a specimen were of a uniform size, then the Griffith energy balance criterion would appear to predict that P is proportional to R^2, in contradiction to Auerbach's empirical law (P∝R). This apparent contradiction was widely studied for some 80 years, and two schools of thought evolved—the flaw statistical explanation and the energy balance explanation.

7.5 Flaw Statistical Explanation of Auerbach's Law

Some workers[8,9] attempted to explain Auerbach's law in terms of the surface flaw statistics of the specimen. It was argued that for a larger indenter radius, the increased chance that the region of maximum tensile stress would encompass a particularly large flaw may result in the formation of a cone crack at a reduced load, thus reducing the R^2 dependency. The main criticisms of the flaw statistical explanation are first that, it is extremely improbable that every piece of material would have the exact flaw distribution required to produce the linear form of Auerbach's law. Second, since smaller indenters sample smaller areas of specimen surface, the scatter in results would be expected to increase with decreasing R. Langitan and Lawn[10] claim that this scatter is not observed, although the data of Hamilton and Rawson[8] appear to show otherwise. Finally, the flaw statistical explanation predicts that if all flaws are of the same size, then P is proportional to R^2 if one applies the Griffith energy balance criterion as given by Eq. 7.4d. Langitan and Lawn[10] show that there does exist a range of flaw sizes for which Auerbach's law still holds, even when all flaws are of the same size.

7.6 Energy Balance Explanation of Auerbach's Law

Despite some quantitative agreement with experimental results, flaw statistical explanations of Auerbach's law were never considered entirely satisfactory. An alternative approach, based upon fracture mechanics principles, was proposed by Frank and Lawn[11] in 1968 and given a more complete treatment by Mouginot and Maugis[12] in 1984.

For the case of a crack in a nonuniform stress field, Frank and Lawn[11] showed that the stress intensity factor may be calculated using the prior stress field along the proposed crack path using:

$$K_1 = \frac{2}{\sqrt{\pi}} \int_0^c c^{1/2} \frac{\sigma(b)}{(c^2 - b^2)^{1/2}} db \qquad (7.6a)$$

where c is the length of the crack and b is a variable which represents the length of travel along the crack path (see Chapter 2). The apparent violation of the Griffith energy balance criterion embodied in Eq. 7.4d is a consequence of the assumption that the stress distribution along the length of a cone crack is uniform, and equal to the surface stresses given by the Hertz equations. In fact, the tensile stress diminishes very quickly with depth into the specimen, and hence the pre-existing tensile stress acting along the full length of a surface flaw cannot be considered uniform over its length. Frank and Lawn determined the stress intensity factor for the special case of a crack path that started at the radius of the circle of contact and followed the σ_3 stress trajectory down into the interior of the specimen in terms of the prior stress field using Eq. 7.6a. The fracture mechanics analysis in the indentation stress field predicted the formation of a hitherto unobserved shallow ring crack which precedes the formation of the more familiar cone crack. They proposed that Auerbach's law relates to the observation of a fully developed cone crack rather than the seminal ring crack. In a further development of this idea, Mouginot and Maugis[12] applied Eq. 7.6a for potential crack paths for a range of starting radii in the vicinity of the indenter and showed that Auerbach's law is a consequence of the interaction between the diminishing stress field, the indenter radius, and the starting radius of the cone crack. They argued that for a high density of flaws of uniform size, the cone crack is initiated at the radius for which the strain energy release rate is greatest. In this latter treatment, the seminal ring crack is not a necessary feature of the analysis.

Now, Eq. 7.6a applies to a straight crack. To account for the change in stress intensity factor for an expanding cone crack in which the width of the crack front increases as the crack path increases, Mouginot and Maugis include a correction in Eq. 7.6a to give:

$$K_1 = \frac{2}{\sqrt{\pi}} \int_0^c c^{1/2} \frac{r_b}{r_c} \frac{\sigma(b)}{(c^2 - b^2)^{1/2}} db \qquad (7.6b)$$

where $2\pi r_c$ represents the length of the crack front at the tip of the cone crack, and $2\pi r_b$ is the crack length at the point defined by the variable b at which $\sigma(b)$ applies. In this form, the integral includes the change in length of crack front as b increases from 0 to c. There are several important assumptions embodied in

122   Chapter 7. Hertzian Fracture

the use of Eqs 7.6a or 7.6b to the Hertzian crack system as presented here. First, Eq. 7.6a, in the strictest sense applies to one crack tip of a fully embedded, double-ended, symmetric crack in an infinite solid. If we assume that the rate of strain energy release for a single ended crack is exactly half that of a double-ended crack, then Eqs. 7.6a and 7.6b apply equally well to the tip of a single-ended crack and thus are appropriate for the present analysis*. Second, although it is customary to apply a +12% correction to the value of K_1 for surface flaws (see Chapter 2), this is only generally applicable to surface flaws being acted on by a uniformly applied stress and cannot be justified in the sharply diminishing indentation stress field. Third, we are assuming that the crack path is defined by the σ_3 stress trajectory, since it has been intuitively assumed that this represents the maximum strain energy release rate. In the case of Hertzian cone cracks, the angle of the cone is generally observed to be somewhat shallower than that of the σ_3 stress trajectory. For example, for Poisson's ratio $\nu = 0.2$, the angle of the Hertzian cone is approximately 22°, whereas the σ_3 stress trajectory makes an angle of about 33° to the specimen surface. Despite this difference, the procedure to be followed here need not be invalidated since the features of interest in the analysis occur within the very beginnings of the evolving crack where the crack path is almost normal to the specimen free surface (i.e., in the seminal ring rather than the fully produced cone). Finally, throughout this discussion, circular symmetry is always assumed. Flaw size refers to the flaw depth, and not its length along the surface. The growth of a flaw into a circular ring is not considered here.

Equation 7.6b may be rewritten with stresses in terms of the mean contact pressure $p_m = P/(\pi a^2)$ and distances expressed with respect to the contact radius a such that:

$$f(b/a) = \frac{\sigma(b/a)}{p_m} \qquad (7.6c)$$

Combining Eqs. 7.6b and 7.6c, we may define a function $\phi(c/a)$, related to K_{IC}, as:

$$\phi(c/a) = \frac{c}{a}\left(\int_0^{c/a} \frac{r_b}{r_c}\left(\frac{c^2}{a^2} - \frac{b^2}{a^2}\right)^{-1/2} f(b/a) \; d(b/a)\right)^2 \qquad (7.6d)$$

Since $p_m = P/\pi a^2$, Eq. 7.6d allows the Griffith criterion at the critical fracture condition for the case of the sphere to be expressed in terms of R as:

---

* It would be instructive to review Section 2.5.2. Equations 2.5.2a and 2.5.2b apply to one end of a double-ended crack and not to a single crack tip in isolation. However, it is not unreasonable to assume that the rate of strain energy release is equal for both crack tips, hence the use of Eq. 2.5.2d for the Hertzian crack in Eq. 8.6a.

$$G = 2\gamma = \frac{3(1-v^2)P}{\pi^3 kR}\phi(c/a) \qquad (7.6e)$$

and for either the sphere or punch in terms of a:

$$G = 2\gamma = \frac{4(1-v^2)P^2}{\pi^3 Ea^3}\phi(c/a). \qquad (7.6f)$$

The function $\phi(c/a)$ contains an integral which is characteristic of the pre-existing stress field. The function $\phi(c/a)$ must be evaluated for a particular starting radius, r_o/a, since this determines the values of the stress along the crack path.

Rearranging Eq. 7.6f gives the critical load for fracture $P = P_c$ for the case of the sphere or the punch:

$$P_c = \left(\frac{a^3}{\phi(c/a)}\right)^{1/2}\left(\frac{\pi^3 E 2\gamma}{4(1-v^2)}\right)^{1/2} \qquad (7.6g)$$

The factors in the second term on the right-hand side of Eq. 7.6g are all material constants. It would appear therefore that Auerbach's empirical law would be consistent with the analysis if $\phi(c/a)$ is also a constant, since the critical load would then be proportional to $a^{3/2}$ (which according to Eq. 7.3b, is equivalent to Auerbach's law). However, $\phi(c/a)$ cannot be assumed constant since it contains terms for the stress field, the initial flaw size and indenter radius—all of which are variables. It is later shown that there is a range of values of stress level, indenter radius, and flaw size for which $\phi(c/a)$ is nearly constant. This range of c_f/a is therefore called the "Auerbach range."

Figure 7.6.1 shows values for σ_1 along the path of the σ_3 stress trajectory for different starting radii for both spherical and flat punch indenters and illustrates the diminishing stress field along the prospective crack paths. It can be seen that there is a range of crack lengths for which the maximum principal stress σ_1 remains at a higher level for cracks that are initiated further away from the indenter than for those that commence nearer to the indenter. This apparently anomalous result occurs because cracks that commence very close to the indenter propagate more quickly away from the surface to where the stresses are much less.

For a surface with a high density of flaws, the largest strain energy release rate for a given flaw size determines where and at what load a flaw will develop into a crack. The integral in Eq. 7.6d may be evaluated numerically for the stress distribution along each σ_3 stress trajectory and plotted as a function of c/a as shown in Fig. 7.6.2. The value of $\phi(c/a)$ for any particular normalized radius r_o/a is proportional to the strain energy release rate for a crack of size c/a that commences at radius r_o/a. For any flaw size c_f, there is a particular radius, r_o, for which the strain energy release rate is greatest. This corresponds to the upper envelope of the curves of $\phi(c/a)$ in Fig. 7.6.2. This upper envelope, not drawn in these figures, is denoted as $\phi(c_f/a)$. When the indenter load is steadily increased, the Griffith criterion will be first met when the strain energy release rate, given by Eqs. 7.6e and 7.6f with the envelope of the curves $\phi(c_f/a)$, becomes equal to twice the fracture surface energy. A cone crack will initiate at the lowest load for which a flaw of size c_f/a exists in the specimen at a radius for which $\phi(c_f/a)$ is greater than the critical value.

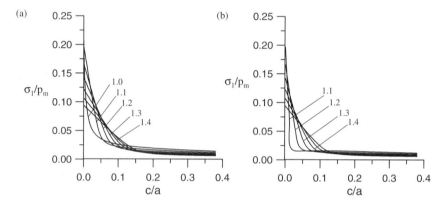

Fig. 7.6.1 Normalized radial stress σ_1/P_m plotted as a function of normalized distance c/a along the σ_3 stress trajectory for different starting radii r_o/a for (a) spherical indenter and (b) cylindrical flat punch indenter. Stresses and trajectories calculated for $\nu = 0.26$ (after reference 3).

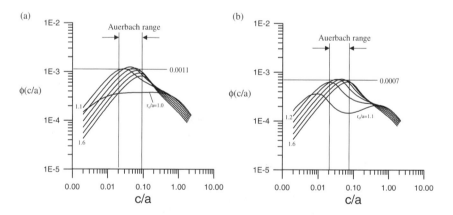

Fig. 7.6.2 Strain energy release function $\phi(c/a)$ as a function of normalized crack length c/a, for different starting radii r_o/a for (a) spherical indenter and (b) cylindrical flat punch indenter. The Auerbach range, where the outer envelope of $\phi(c/a)$ is approximately constant, is indicated in each figure along with the estimated value of ϕ_a (after reference 3).

For a high density of very small flaws, in the size range $c_f/a < 0.01$, the critical load P_c, given by Eq. 7.6g, decreases as the flaw size increases, since the stress level along the length of the flaw is fairly constant and is approximately equal to the surface stress as given by the Hertz equation. In this case, the Griffith criterion for a uniform constant stress level may be employed. Smaller flaws are more likely to extend at a lower r_o/a, since the surface stress level is higher closer to the contact radius. Auerbach's law would not hold in this case.

For larger flaws, in the size range $0.1 < c_f/a < 0.2$, the situation is qualitatively different. Equation 7.6g and Fig. 7.6.2 show that the critical load increases with increasing flaw size because the strain energy release rate given by $\phi(c/a)$ decreases with increasing flaw size. The reason for this surprising result is in the form of the integral in Eq. 7.6d. The strain energy release rate depends on both the stress distribution along the flaw and the factor $(c^2-b^2)^{-1/2}$. Larger values of c cause the integral to evaluate to a lower value compared to smaller flaws at the same r_o.

From Eq. 7.6g, $P_c/a^{3/2}$ is proportional to $\phi(c_f/a)^{-1/2}$. Figure 7.6.2 shows that there is a range of c_f/a where the outer envelope, $\phi(c_f/a)$, (and hence $\phi(c_f/a)^{-1/2}$) is fairly constant. *This is the Auerbach range.* In this range, the critical load P_c which initiates fracture is virtually independent of the flaw size and is therefore proportional to $a^{3/2}$. Assuming the existence of flaws of all sizes everywhere on the specimen surface, then for a particular flaw size, the starting radius is that which gives the maximum strain energy release rate. The Griffith criterion will be first met, upon increasing load, at the position where the maximum strain energy release rate occurs. For another flaw size, the starting radius is different but the strain energy release rate, and hence the critical load, is not much different.

For flaws within the Auerbach range of flaw sizes, the minimum critical load is given the symbol P_a and is found from $\phi(c/a) = \phi_a$ and Eq. 7.6g:

$$P_a = \left[\frac{2k\pi^3\gamma}{(1-v^2)\beta\phi_a}\right] R \tag{7.6h}$$

for the sphere and:

$$P_a = \left[\frac{E\pi^3\gamma}{(1-v^2)2\phi_a}\right]^{1/2} a^{3/2} \tag{7.6i}$$

for the punch where the term in the square bracket in Eq. 7.6h is the Auerbach constant directly. In Eqs. 7.6h and 7.6i, ϕ_a is the value of $\phi(c/a)$ at the plateau. From Fig. 7.6.2, this is estimated to be at $\phi(c/a) = 0.0011$ for the case of the sphere and $\phi(c/a) = 0.0007$ for the punch. The value of ϕ_a is important since it influences the fracture surface energy, which is estimated from data obtained from indentation experiments. Combining Eqs. 7.6h and 7.6i, and 7.6e and 7.6f, it may be shown that:

$$\frac{G}{2\gamma} = \left(\frac{P}{P_a}\right) \frac{\phi(c/a)}{\phi_a} \qquad (7.6j)$$

for the sphere and:

$$\frac{G}{2\gamma} = \left(\frac{P}{P_a}\right)^2 \frac{\phi(c/a)}{\phi_a} \qquad (7.6k)$$

for the case of the punch. Plots of $G/2\gamma$ as calculated using Eqs. 7.6j and 7.6k are shown in Fig. 7.7.1.

The term "fracture" in the present context signifies the extension of a flaw to a circular ring crack concentric with the contact radius. Once a flaw has become a propagating crack, it extends according to the strain energy release function curve, Fig. 7.6.3, appropriate to its starting radius. The development of this starting flaw into a ring crack precludes the extension of other flaws in the vicinity, even though the value of $\phi(c/a)$ for those flaws at some applied load above the flaw initiation load may be larger than that calculated for the starting flaw as it follows its $\phi(c/a)$ curve. This is because the conditions that determine crack growth depend on the *prior* stress field. The function $\phi(c/a)$ can be used to describe the initiation of crack growth for all flaws that exist in the prior stress field but can only be considered applicable for the subsequent elongation for the flaw that actually first extends. Note that the Auerbach range shown in Fig. 7.6.2 corresponds to crack lengths c/a in the range 0.01 to 0.1. These crack lengths correspond to the initial ring shape of the crack and not the developing cone, and hence the difference in observed angle of cone crack and σ_3 stress trajectory mentioned previously does not alter the results or the validity of the method of analysis presented here.

The energy balance explanation of Auerbach's law requires the existence of surface flaws within the so-called "Auerbach range." We are now in a position to develop a procedure to determine the conditions for the initiation of a Hertzian cone crack in specimens whose surfaces contain flaw distributions of a specific character. This procedure brings together the flaw statistical and energy balance explanations of Auerbach's law[3,13].

7.7 The Probability of Hertzian Fracture

7.7.1 Weibull statistics

Both the size and distribution of surface flaws characterize the strength of brittle solids and the probability of failure of a specimen of surface area A subjected to a uniform tensile stress σ can be calculated using Weibull statistics[14] (see Chapter 4):

$$P_f = 1 - \exp\left(-kA\sigma^m\right) \qquad (7.7.1a)$$

where m and k are the Weibull parameters. The parameter m describes the spread in strengths (a large value indicating a narrow range), and the parameter k is associated with the "reference strength" and the surface flaw density of the specimen. Typical values for as-received soda-lime glass windows are[15] m = 7.3 and k = 5.1×10^{-57} m^{-2} Pa$^{-7.3}$.

The probability of failure given by Eq. 7.7.1a is equal to the probability of finding a flaw within an area A of the specimen surface that is larger than the critical flaw size (as given by the Griffith criterion) for a uniform stress σ. The critical flaw size is given by Eq. 7.4c with $K_1 = K_{1C}$.

On the surface of any given specimen, there may exist a considerable number of flaws of lengths below, above, and within the Auerbach range on the surface of a specimen. The probability of failure (initiation of a Hertzian cone crack) for a given indenter load depends directly on the probability of finding a surface flaw of critical size within the indentation stress field. Critical stress and flaw size are related by Eq. 7.4c, where the stress is applied along the full depth of the flaw. In an indentation stress field, however, this only applies for very small flaws where the tensile stress is given by Hertz's equations. The uniform stress field approximation gets progressively worse as the Auerbach range is approached. For larger flaws, within the Auerbach range, the fracture load becomes nearly independent of the flaw size since the maximum strain energy release rate, as described by the outer envelope of the curves of Fig. 7.6.2, is approximately constant. The probability of fracture from these flaws must therefore be expressed in terms of the probability of finding a flaw of the required size at a starting radius commensurate with the curves of Fig. 7.6.2.

7.7.2 Application to indentation stress field

We are now in a position to calculate the probability of fracture for a given load and radius of indenter. Let P_a be the minimum critical load for values of c/a within the Auerbach range. Figure 7.7.1 shows the relationship between the normalized strain energy release rate G/2γ, flaw size c/a, and starting radius r_o/a for three different values of P: P−, a load below the minimum critical load; P_a, the minimum critical load; and P+, a load greater than the minimum critical load. The Griffith criterion is met when G/2γ ≥ 1. On this diagram, the line G/2γ = 1 has been drawn at positions corresponding to P−, P_a and P+. This allows the graph to be presented more clearly, showing only one family of curves. The curves shown in Fig. 7.7.1 rely only upon the choice of ϕ_a and are independent of the value of γ. However, if one wishes to draw curves as in Fig. 7.7.1 for a particular indenter load, then P_a must be determined from Eq. 7.6.k or Eq. 7.6.l, for frictionless contact, which requires an estimate of γ.

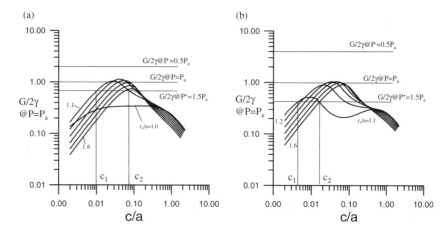

Fig 7.7.1 Relationship between strain energy release rate, G and flaw size c/a for different indenter loads P/P_a. The vertical axis scaling applies to $P/P_a = 1$. The vertical axis positions for the condition $G = 2\gamma$ for different ratios P/P_a are drawn relative to the family of curves shown. The flaw size range for $G/2\gamma > 1$ for a starting radius $r_o/a = 1.2$ for $P/P_a = 1.5$ is indicated (after reference 3).

It is immediately evident that if the load is less than the minimum critical load P_a, failure will not occur from any flaws, no matter how large, since the Griffith criterion is never met. It can be seen that failure can only occur from flaws within the Auerbach range for loads equal to or greater than P_a. Fracture from flaws of size below, including, and beyond the Auerbach range can only occur if the load is greater than P_a. At a load P+, greater than P_a, the Griffith criterion is met for various ranges of flaw sizes which depend on the particular values of starting radii. Fracture will occur from a flaw located at a particular starting radius if that flaw is within the range for which $G/2\gamma \geq 1$ for that radius.

This range of flaw sizes can be determined from Fig. 7.7.1 and is given by the c/a axis coordinates for the upper and lower bounds of the region where $G/2\gamma > 1$ for the curve that corresponds to the radius under consideration. The problem has been reduced to that from calculating the probability of indentation fracture occurring at a particular radius and load to the probability of finding at least one flaw within a specific size range at that radius. For the case of a punch, the procedure is straightforward since the radius of circle of contact a is a constant. For a sphere, the contact radius depends on the load, and the procedure for determining the required flaw sizes is slightly more complicated.

To determine these probabilities, it is convenient to divide the area surrounding the indenter into n annular regions of radii r_i (i = 1 to n). To determine the probability of finding a flaw that meets the Griffith criterion within each annular region, Eq. 7.7.1a may be used. Eq. 7.7.1a gives the probability of failure for an applied uniform stress but also can be used to calculate the probability of finding a flaw of size greater than or equal to the critical value for that stress, as given by Eq. 7.4c, within an area A of the surface of the solid. The strength parameters, m and k, for Eq. 7.7.1a are those appropriate to the specimen surface condition. The probabilities calculated for each annular region can be suitably combined to yield a total probability of failure for a particular indenter load and radius for a given surface flaw distribution.

We proceed as follows. Curves as shown in Fig. 7.7.1 are drawn for a particular value of indenter load P. Consider an annular region with radius r_i and area δA_i. The range of values of flaw size that satisfies the Griffith criterion may be determined for this region by considering the appropriate line for $\phi(c/a)$ in Fig. 7.7.1. For example, the vertical lines in Fig. 7.7.1 show the range of flaw sizes for $P/P_a = 1.5$, which, should they exist within the increment centered on $r_i/a = 1.1$, will cause fracture at that radius. Let this range be denoted by $c_1 \le c \le c_2$. We therefore require the probability of finding such a flaw within this size range in the area δA. This is equal to the difference between the probability of finding a flaw of size $c > c_1$ and the probability of finding a flaw of size $c > c_2$. However, the probability of finding a flaw of size greater than a specific size, say c_1, within the area δA_i is precisely equal to the Weibull probability of failure (Eq. 7.7.1a) under the corresponding critical stress as given by Eq. 7.4c.

Once a particular indenter size has been specified, the probability of finding a flaw of size greater than c_1 within the annular region of radius r_i and width δr_i, which has an area $\delta A_i = 2\pi r_i \delta r_i$, is:

$$P_i(c > c_1) = 1 - \exp\left[-k 2\pi r_i \delta r_i \left(\frac{K_{1C}}{(\pi c_1)^{1/2}}\right)^m\right] \qquad (7.7.2a)$$

Similarly, the probability of finding a flaw of size greater than c_2 within the same area element δA_i is given by:

$$P_i(c > c_2) = 1 - \exp\left[-k 2\pi r_i \delta r_i \left(\frac{K_{1C}}{(\pi c_2)^{1/2}}\right)^m\right] \qquad (7.7.2b)$$

The probability of finding a flaw of size in the range $c_1 \le c \le c_2$ within area δA_i is the difference in probabilities given by Eqs. 7.7.2a and 7.7.2b and is equal to the probability of failure from a flaw of size within that range.

$$P_{fi}(c_1 \le c \le c_2) = P_i(c > c_1) - P_i(c > c_2) \qquad (7.7.2c)$$

The values c_1 and c_2 may be determined for all annular regions by inspection of Fig. 7.7.1. Since a two-parameter Weibull function gives a nonzero probability of failure for even the lowest stresses, it would appear that the upper limit of r_i/a should extend to the full dimensions of the specimen, where the effect of the indentation stress field may still be apparent. However, if one is interested in

130   Chapter 7. Hertzian Fracture

loads near to the minimum critical load for flaws within the Auerbach range, P_a, then it is necessary to consider only starting radii that correspond to the upper end of the Auerbach range; that is, $r_i/a = 1.5$, which gives a maximum $\phi(c_f/a)$ at $c/a = 0.1$. The probability of fracture not occurring from a flaw within the region δA is found from:

$$P_{s_i} = 1 - P_{f_i} \qquad (7.7.2d)$$

The probability of survival for the entire region of n annular elements surrounding the indenter is thus given by:

$$P_S = P_{s_1} P_{s_2} P_{s_3} ... P_{s_i} ... P_{s_n} \qquad (7.7.2e)$$

Therefore, finally, the probability of failure P_F for the entire region, at the load P/P_a, is then given by:

$$P_F = 1 - P_S \qquad (7.7.2f)$$

This calculation is repeated for different values of P/P_a to obtain the dependence on indenter load of probability of failures for a particular value of indenter radius. For the case of a sphere, the situation is complicated by the expanding radius of circle of contact with increasing load. Combining Eqs. 7.2a and 7.6h, it is easy to show that the radius of circle of contact for a given radius of indenter and ratio P/P_a may be calculated from:

$$a^3 = \frac{8}{9} \frac{k^2}{E} \frac{\gamma \pi^3}{(1-v^2)\phi_a} \frac{P}{P_a} R^2 \qquad (7.7.2g)$$

This permits values for c_f to be determined as a function of P/P_a for a constant R and proceeding as for the case of the punch. Figure 7.7.2 shows the probability of failure as a function of indenter load for a particular size of indenter for both spherical and cylindrical punch indenters.

Calculated values are shown along with those determined from indentation experiments. The experimental work was performed on as-received soda-lime glass specimens using a hardened steel cylindrical punch and a tungsten carbide sphere. Agreement is fairly good especially when one considers that the Weibull parameters used in the calculations were determined on glass specimens from a completely different source than those used in the experimental work. The curves in Fig. 7.7.2 rely on an estimation of the fracture surface energy γ in Eqs. 7.6h and 7.6i.

Although the fracture surface energy may in principle be determined from indentation tests, such estimations are inaccurate due to the inevitable presence of friction between the indenter and the specimen. Nevertheless, the calculated curves in Fig. 7.7.2 have been obtained using Eqs. 7.6h and 7.6i with fracture surface energies determined from the experimental data (see Section 7.8). In Fig. 7.7.2, the cutoff at P_a for each indenter size indicates a zero probability of failure for loads below the minimum critical load.

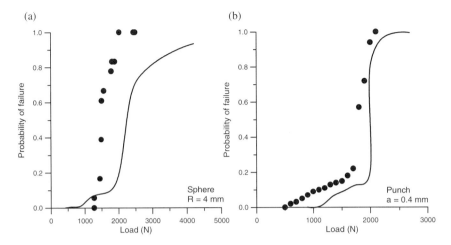

Fig. 7.7.2 Probability of failure versus indenter load for as-received soda-lime glass for (a) spherical indenter R = 4 mm and (b) cylindrical flat punch indenter a = 0.4 mm. Solid line indicates calculated values with surface energy γ as given in Table 7.1 and (●) indicates experimental results (after reference 3).

It is of interest to note that the probability of indentation failure may be expressed in terms of Weibull strength parameters that are usually determined from bending tests involving a stress field which is nearly constant with depth over a distance characteristic of the flaw size. This is possible since the probability of indentation failure is being expressed in terms of the probability that certain areas of surface contain flaws within various size ranges. This probability is a property of the surface, and the surface strength parameters m and k may be determined through bending tests. A suitable combination of these probabilities gives the probability of failure for the special case of the diminishing stress field associated with an indentation fracture.

7.8 Fracture Surface Energy and the Auerbach Constant

7.8.1 Minimum critical load

The procedure given in previous sections for calculating the probability of initiation of a Hertzian cone crack relies on an estimation of the fracture surface energy of the specimen material. Experimental indentation work reported by Fischer-Cripps and Collins[13] indicates a fracture surface energy nearly 2.5 times that determined by other means[7], causing those workers to postulate that the inevitable presence of friction beneath the indenter leads to an increase in the apparent surface energy estimated from indentation experiments, even with lubricated contacts. Estimations of fracture surface energy are best undertaken with respect to the minimum critical load for failure.

As before, let P_a denote the minimum critical load for an indentation fracture to occur. We would expect this minimum critical load to correspond to the fracture load observed in experiments on glass with a high density of flaws (i.e., on abraded glass). Equations 7.6h and 7.6i predict a straight line relationship between spherical indenter radius and the punch radius to the 3/2 power, respectively and the *minimum critical load*. This is expected since Eqs. 7.6h and 7.6i assume a specimen surface containing flaws of *all sizes* and do not give any information about the probability of finding a particular sized flaw at a particular starting radius. As the indenter size is increased, the flaw size corresponding to the Auerbach range also increases and it is from flaws within the Auerbach range that failure first occurs since the functions $\phi(c/a)$, as shown in Figs. 7.6.2, are a maximum in the Auerbach range of flaw sizes. From Eq. 7.6h, the Auerbach constant is given by:

$$A = \left[\frac{2k\pi^3\gamma}{(1-\nu^2)3\phi_a}\right] \quad (7.8.1a)$$

For the case of the punch, the Auerbach constant for an "equivalent" sphere of radius R giving a contact circle of radius a may be found from Eq. 7.6i and the Hertz equation, Eq. 7.2a:

$$A = \left[\frac{E\pi^3\gamma}{(1-\nu^2)2\phi_a}\right]\left(\frac{4k}{3E}\right) = \left[\frac{2k\pi^3\gamma}{(1-\nu^2)3\phi_a}\right] \quad (7.8.1b)$$

As can be seen from Eqs. 7.8.1a and 7.8.1b, the Auerbach constant depends upon the value of fracture surface energy γ.

Figure 7.8.1 shows experimental data for the minimum critical load obtained on abraded soda-lime glass using both spherical and flat punch indenters. The data for the punch have been plotted as a function of $a^{3/2}$ to give a linear relationship with the minimum critical load; the actual punch diameter is indicated for each data point. The slope of the line of best fit (solid lines in Figs. 7.8.1) through the data provides an estimate of the magnitude of the Auerbach constant A with ϕ_a estimated from the plateau regions of Fig. 7.6.2. Values of surface energy γ can then be calculated from Eqs. 7.8.1a and 7.8.1b.

Values of A and γ estimated in this manner are given in Table 7.1. As can be seen, the fracture surface energies obtained using this method for the two indenters are not all that different, although they are appreciably higher (by a factor of $\approx$2) than the expected value of $\gamma = 3.5$ J/m^2 for this material. Differences between the value of A obtained from the experiments using the sphere and that with the punch are most probably due to the different dependence on friction on the indentation response of the two types of indenter. It should be noted that the probability of failure shown in Fig. 7.7.2 has been calculated using the fracture surface energies shown in Table 7.1.

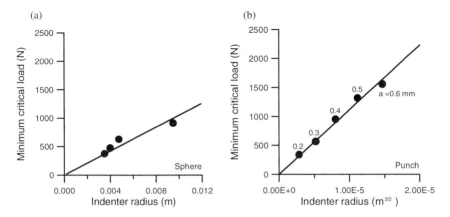

Fig. 7.8.1 Minimum critical load versus indenter radius for (a) spherical and (b) cylindrical indenters for abraded glass. (●) indicates experimental results with lubricated contacts. The horizontal axis in (b) is given as the indenter radius raised to the 2/3 power, the actual radius of the indenter in mm is shown for each experimental result. The solid line is the best linear fit through the experimental data, the slope of which is used to determine the value of the Auerbach constant in Table 7.1 (after reference 3).

Table 7.1 Fracture surface energy and Auerbach constant for soda-lime glass from indentation tests with spherical and cylindrical flat punch indenters.

	Sphere	Punch
Fracture surface energy (J/m^2)	8.88	7.46
Auerbach constant (N/m)	10.5×10^4	13.8×10^4

Environmental effects also have an influence on the probability of fracture, and an equivalent load may be calculated using the "Modified crack growth model" presented in Chapter 3.

7.8.2 Median fracture load

In an attempt to explain Auerbach's law, some workers have correlated the values of scatter in the fracture loads with the surface flaw characteristics of the specimen to arrive at a relationship between the median fracture load and indenter radius. For example, Oh and Finnie[9] initially determined Weibull parameters from bending tests on glass strips. The probabilities of failure for annular regions surrounding the indenter were calculated on the basis of a nondiminishing stress field and combined to give a total probability of failure.

From these results, the expected value of the fracture load for a given indenter size was calculated and compared with the mean fracture load obtained from indentation experiments. In a similar series of experiments, Hamilton and Rawson[8] determined the Weibull parameters that best described indentation fractures. Argon[16] determined a strength distribution function that described the variation in fracture load for a fixed indenter radius, but he did not express his results in terms of Weibull strength parameters.

Here, no distinction is made between the mean load and the median fracture load, although it should be noted that the median fracture load corresponds to a probability of failure of precisely 50%. The mean fracture load may thus be estimated by determining the load for P_f = 50% in Fig. 7.7.2. Estimates for both the sphere and the punch are plotted in Fig. 7.8.2 and compared with those determined from experiments on as-received glass. Although the theory predicts that, within the Auerbach range there is a linear relationship between the *minimum critical load* and the indenter radius, there is no particular reason why this should be so for *median* or *mean fracture loads*. Indeed, if a linear relationship existed, it would be expected that the Auerbach constants obtained from such data would be largely determined by the flaw statistics of the sample rather than by intrinsic material properties.

Figure 7.8.2 shows that a linear relationship is not indicated for the mean fracture load for both spherical and punch indenters, in either calculated or experimental determinations.

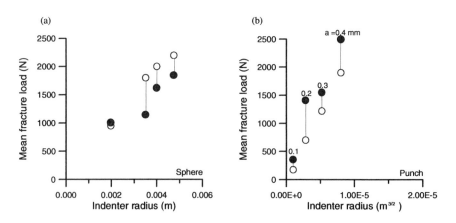

Fig. 7.8.2 Mean failure load versus indenter radius for (a) spherical and (b) cylindrical indenters for as-received glass. The horizontal axis in (b) is given as the indenter radius raised to the 2/3 power; the actual radius of the indenter in mm is shown for each experimental result. (O) indicates calculated values with surface energy γ as given in Table 7.1, and (●) indicates experimental results with lubricated contacts.

7.9 Cone Cracks

7.9.1 Crack path

Despite the apparent satisfaction at being able to account for the initiation of a Hertzian cone crack in terms of both flaw statistics and the requirements of the Griffith energy-balance criterion, there exists an important unexplained anomaly in this type of treatment. The method relies on a calculation of stress intensity factor along an assumed crack path which is assumed to be normal to the maximum principal tensile stress, (i.e., the σ_3 stress trajectory) of the prior stress field. However, in many materials, there is a disparity between the path delineated by this calculated stress trajectory and that taken by the conical portion of an actual crack. Calculations show that, in glass with Poisson's ratio = 0.21, the angle of the cone crack, if it were to follow a direction given by the σ_3 trajectory, should be approximately 33° to the specimen surface. The actual angle is dependent on Poisson's ratio. However, experimental evidence[17] is that the angle is typically much shallower, by up to 10° in some cases. Lawn, Wilshaw, and Hartley[18] attempted to resolve this disparity by analytical computation but were unsuccessful. Until recently, no one has attended to this issue, except perhaps for Yoffe[19], who predicted that the answer lies in the modification to the pre-existing stress field by the presence of the actual cone crack as it progressed through the solid, and Lawn[20], who proposed a change in local elastic properties in the vicinity of the highly stressed crack tip.

Although the notion that the path of a crack may be predicted in terms of the pre-existing stress field may appear to be untenable, it is fundamentally correct as long as the crack proceeds normal to the maximum pre-existing principal tensile stress σ_1 (i.e., tensile, or Mode I fracture), since it is this path that usually results in a maximum value for stress intensity factor K and hence the greatest value of energy release G. In the Hertzian stress field, this path would be the trajectory of the minimum principal stress σ_3. Recently, Kocer and Collins[21] have used a numerical approach to show that the trajectory of a Hertzian cone crack is such that incremental growth according to a criterion of maximum strain energy release occurs at a shallower angle than would be expected from a consideration of just the σ_3 stress trajectory. This numerical result is in complete agreement with experimental evidence and thus infers that the path of maximum release of strain energy is not that resulting in pure Mode 1 loading. This is an extremely important result, not only for the Hertzian cone crack system but for any crack system involving nonuniform triaxial stress fields. However, despite the satisfaction of knowing that the crack path does represent the path of maximum G, the reason for the disagreement with the stress trajectory of the prior field is not yet explained. No doubt, the presence of the specimen free surface has an effect (the Green's function approach of Eq. 7.6d is for an embedded crack in an infinite solid), as does the presence of interfacial friction between the indenter and specimen (which modifies the elastic stress field calculated using the equations of Chapter 5).

7.9.2 Crack size

Once a cone crack has initiated, the rate of strain energy release will follow the appropriate curve of Fig. 7.7.1 for the particular value of c/a at which initiation occurred. Note that the rising portion of these curves represents unstable crack growth. The falling edge of any one curve is stable crack growth and the crack will assume an equilibrium length when G/2γ approaches 1. Increasing the load shifts all these curves upward and the crack extends in a stable manner until a new equilibrium is reached. Experiments and theoretical analysis[22,23] show that the radius of the stabilized cone crack varies as the indenter load P raised to the 2/3 power.

$$\frac{P^2}{R^3} = 2\gamma ED \qquad (7.9.1)$$

where D is a constant dependent on Poisson's ratio and R is the radius of the base of the fully formed cone crack. Roesler determined $D = 2.75 \times 10^{-3}$ for $\nu = 0.25$. Equation 7.9.1 is Roesler's "scaling" law and is obtained from a fundamental analysis of the strain energy contained within the truncated cone geometry of the Hertzian cone crack system.

References

1. H. Hertz, "On the contact of elastic solids," J. Reine Angew. Math. 92, 1881, pp. 156–171. Translated and reprinted in English in *Hertz's Miscellaneous Papers*, Macmillan & Co., London, 1896, Ch. 5.
2. H. Hertz, "On hardness," Verh. Ver. Beförderung Gewerbe Fleisses 61, 1882, p. 410. Translated and reprinted in English in *Hertz's Miscellaneous Papers*, Macmillan & Co, London, 1896, Ch. 6.
3. A.C. Fischer-Cripps, "Predicting Hertzian fracture," J. Mater. Sci., 32 5, 1997, pp. 1277–1285.
4. F. Auerbach, "Measurement of hardness," Ann. Phys. Leipzig, 43, 1891, pp. 61–100.
5. A.A. Griffith, "Phenomena of rupture and flow in solids," Philos. Trans. R. Soc. London, Ser. A221, 1920, pp. 163–198.
6. G.R. Irwin, "Fracture" in *Handbuch der Physik*, , 6, Springer-Verlag, Berlin, 1958, p. 551.
7. S.W. Freiman, T.L. Baker and J.B. Wachtman, Jr., "Fracture mechanics parameters for glasses: a compilation and correlation." in *Strength of Inorganic Glass*, edited by C.R. Kurkjian, Plenum Press, New York, 1985, pp. 597–607.

8. B. Hamilton and R. Rawson, "The determination of the flaw distributions on various glass surfaces from Hertz fracture experiments," J. Mech. Phys. Solids 18, 1970, pp. 127–147.
9. H.L. Oh and I. Finnie, "The ring cracking of glass by spherical indenters," J. Mech. Phys. Solids 15, 1967, pp. 401–411.
10. F.B. Langitan and B.R. Lawn, "Hertzian fracture experiments on abraded glass surfaces as definitive evidence for an energy balance explanation of Auerbach's law," J. Appl. Phys. 40 10, 1969, pp. 4009–4017.
11. F.C. Frank and B.R. Lawn, "On the theory of Hertzian fracture," Proc. R. Soc. London, Ser. A229, 1967, pp. 291–306.
12. R. Mouginot and D. Maugis, "Fracture indentation beneath flat and spherical punches," J. Mater. Sci. 20, 1985, pp. 4354–4376.
13. A.C. Fischer-Cripps and R.E. Collins, "The probability of Hertzian fracture," J. Mater. Sci. 29, 1994, pp. 2216–2230.
14. W. Weibull, "A statistical theory of the strength of materials," Ingeniorsvetenskapsakademinshandlingar 151, 1939.
15. W.G. Brown, "A Load Duration Theory for Glass Design," National Research Council of Canada, Division of Building Research, NRCC 12354, Ottawa, Ontario, Canada, 1972.
16. A.S. Argon, "Distribution of cracks on glass surfaces," Proc. R. Soc. London, Ser. A250, 1959, pp. 482–492.
17. R. Warren, "Measurement of the fracture properties of brittle solids by Hertzian indentation," Acta Metall. 26, 1978, pp. 1759–1769.
18. B.R. Lawn, T.R. Wilshaw, and N.E.W. Hartley, "A computer simulation study of hertzian cone crack growth," Int. J. Fract. 10 1, 1974, pp. 1–16.
19. E.H. Yoffe, "Elastic stress fields caused by indenting brittle materials," Philos. Mag. A, 46 4, 1982, p. 617.
20. B.R. Lawn, private communication, 1994.
21. C. Kocer and R.E. Collins, "The angle of Hertzian cone cracks," J. Am. Ceram. Soc., 81 7, 1998, pp. 1736–1742.
22. J.P.A. Tillet, "Fracture of glass by spherical indenters," Proc. Phys. Soc. London, Sect. B69, 1956, pp. 47–54.
23. F.C. Roesler, "Brittle fractures near equilibrium," Proc. Phys. Soc. London, Sect. B69, 1956, pp. 981–982.

Chapter 8
Elastic-Plastic Indentation Stress Fields

8.1 Introduction

In Chapters 5 and 6, we considered the elastic stress fields associated with the contact of elastic solids for indenters of various shapes. It was noted that the indentation stress field could be derived analytically by the superposition of stress fields for a series of point loads arranged to give the required contact pressure distribution for the type of indenter being considered. In the present chapter, we focus on another very important type of indentation, that which occurs with plastic deformation of the specimen material. In brittle materials, this most commonly occurs with pointed indenters such as the Vickers diamond pyramid. In ductile materials, plasticity may be readily induced with a "blunt" indenter such as a sphere or cylindrical punch. Indentation tests are used routinely in the measurement of hardness of materials, but Vickers, Berkovich, and Knoop diamond indenters may be used to investigate other mechanical properties of solids such as specimen strength, fracture toughness, and internal residual stresses. Analysis of the stress fields associated with an elastic-plastic contact is complicated by the presence of plastic deformation in the specimen material. The plastically deformed material modifies the previously described elastic stress fields, and in brittle materials the initiation and growth of cracks often occurs within the specimen on both loading and unloading of the indenter.

8.2 Pointed Indenters

8.2.1 Indentation stress field

In practice, an indentation made with a Vickers pyramidal indenter is initially elastic, due to the finite radius of the indenter tip, but very quickly induces plasticity in the specimen material with increasing load. Removal of load generally results in a residual impression in the surface of the specimen. The elastic stress field is similar to that described in Chapter 5 for a conical indenter, although the four-sided nature of the pyramidal indenter means that the loading is no longer axis-symmetric. However, the general characteristics of the field remain unchanged (more so with increasing distance from the indentation in accordance

Chapter 8. Elastic-Plastic Indentation Stress Fields

with Saint-Venant's principle). With reference to Fig. 5.3.2, it can be seen that σ_1 is tensile, with a maximum value at the specimen surface at r = 0, normal stress σ_3 is compressive, and the hoop stress σ_2 is compressive at the surface and tensile beneath the indentation with a maximum tension in a zone directly beneath the indenter. In these types of indentations, it is the condition of plasticity beneath the indenter that is of considerable practical interest since the modifications to the elastic stress field are responsible for cracking within the specimen for both loading and unloading of the indenter.

Theoretical analysis of the elastic-plastic indentation stress field associated with a pyramidal indenter is difficult, if not impossible, due to the complexities of the nature of the plastic deformation within the specimen material. Since plastic strains in these types of indentations are very much larger than any of the elastic strains, the specimen is usually held to behave as a rigid-plastic material in which plastic flow is assumed to be governed by flow velocity considerations. Marsh[1] compared the plastic deformation in the specimen beneath the indenter to that occurring during the radial expansion of a spherical cavity subjected to internal pressure, an analysis of which was given previously by Hill[2]. The most widely accepted analytical treatment is that of Johnson[3], who replaced the expansion of the cavity with that of an incompressible hemispherical core of material subjected to an internal pressure, the so-called "expanding cavity" model. Analytical models of the elastic-plastic stress field have been proposed by Chiang, Marshall, and Evans[4,5] and also Yoffe[6]. These analyses build on the expanding cavity model of Johnson and include the influence of the specimen free surface. Chiang, Marshall and Evans[4,5] show that the value of the hoop stress in the elastically strained material, on the axis of symmetry, is a maximum at the elastic-plastic boundary and is approximately 0.1–0.2 times the mean contact pressure (p_m), some 10 times that for an equivalent elastic contact with a spherical indenter loaded to the same value of p_m. This result was obtained for analysis of both (axis-symmetric) conical and spherical indenters.

The analytical models mentioned above deal with an axis-symmetric loading of an infinite half-space with a pointed indenter. The most accessible is the Yoffe model[6], which gives the stress distribution outside a hemispherical plastic zone of radius a equal to the radius of the circle of contact (Fig. 8.2.1).

With a coordinate system as shown in Fig. 8.2.2, the stresses are given by:

$$\sigma_r = \frac{P}{2\pi r^2}\left(1-2v-2(2-v)\cos\theta\right)+\frac{B}{r^3}4\left((5-v)\cos^2\theta-(2-v)\right) \quad (8.2.1a)$$

$$\sigma_\theta = \frac{P}{2\pi r^2}\frac{(1-2v)\cos^2\theta}{(1+\cos\theta)}-\frac{B}{r^3}2(1-2v)\cos^2\theta \quad (8.2.1b)$$

$$\sigma_\phi = \frac{P(1-2v)}{2\pi r^2}\left(\cos\theta-\frac{1}{1+\cos\theta}\right)+\frac{B}{r^3}2(1-2v)\left(2-3\cos^2\theta\right) \quad (8.2.1c)$$

$$\tau_{r\theta} = \frac{P(1-2\nu)}{2\pi r^2} \frac{\sin\theta\cos\theta}{1+\cos\theta} + \frac{B}{r^3} 4(1+\nu)\sin\theta\cos\theta \qquad (8.2.1d)$$

$$\tau_{r\phi} = \tau_{\theta\phi} = 0$$

where P is the indenter load and B is a constant which characterizes the extent of the plastic zone. For porous materials beneath a large-angled indenter, the B is small. Yoffe determined that $B = 0.06 p_m a^3$ for soda-lime glass $\nu = 0.26$ with a cone semi-angle of 70° (p_m is the mean contact pressure and a is the radius of circle of contact). Substituting this value of B into Eqs. 8.2.1a to 8.2.1d and normalizing to mean contact pressure p_m and radius of circle of contact a, we obtain:

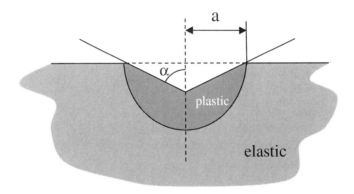

Fig. 8.2.1 Geometry of plastic zone for axis-symmetric conical indenter of semiangle α. It is assumed that the plastic zone meets the surface at r = a.

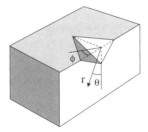

Fig. 8.2.2 Coordinate system and schematic of indentation with pointed indenter.

$$\frac{\sigma_r}{p_m} = \frac{1}{2}\left[\frac{a}{r}\right]^2 (1-2v-2(2-v)\cos\theta)$$
$$+ 0.06\left[\frac{a}{r}\right]^3 4((5-v)\cos^2\theta - (2-v)) \tag{8.2.1e}$$

$$\frac{\sigma_\theta}{p_m} = \frac{1}{2}\left[\frac{a}{r}\right]^2 \frac{(1-2v)\cos^2\theta}{(1+\cos\theta)} - 0.06\left[\frac{a}{r}\right]^3 2(1-2v)\cos^2\theta \tag{8.2.1f}$$

$$\frac{\sigma_\phi}{p_m} = \frac{(1-2v)}{2}\left[\frac{a}{r}\right]^2 \left(\cos\theta - \frac{1}{1+\cos\theta}\right)$$
$$+ 0.06\left[\frac{a}{r}\right]^3 2(1-2v)(2-3\cos^2\theta) \tag{8.2.1g}$$

$$\frac{\tau_{r\theta}}{p_m} = \frac{(1-2v)}{2}\left[\frac{a}{r}\right]^2 \frac{\sin\theta\cos\theta}{1+\cos\theta} + 0.06\left[\frac{a}{r}\right]^3 4(1+v)\sin\theta\cos\theta \tag{8.2.1h}$$
$$\tau_{r\phi} = \tau_{\theta\phi} = 0$$

Equations 8.2.1a to 8.2.1h are obtained by adding the elastic Boussinesq field (the first term in these equations with a $1/r^2$ dependency) to that of a plastic "blister" field (second term with a $1/r^3$ dependency). Yoffe has obtained this blister field by combining a symmetrical center of pressure (similar to that used in "expanding cavity" models to be discussed in Chapter 9) with surface forces which account for the specimen free surface.

Cook and Pharr[7], in a review of the field in 1990, expressed the parameter B in terms of indentation parameters and material properties. They argued that B could be found from:

$$B = 0.0816 \frac{E}{\pi} f\left(\frac{P}{H}\right)^{\frac{3}{2}} \tag{8.2.1i}$$

with f, a densification factor* which varies from 0 (volume accommodated entirely by densification) and 1 (no densification of specimen material), and H the hardness value defined as:

$$H = \frac{P}{2a^2} \tag{8.2.1j}$$

---

* The volume of material displaced by the indenter may be taken up by varying proportions of densification (i.e., compaction), elastic strains, plastic strains, and piling up at surface, etc.

where a is the radius of the circle of contact. Inserting Eq. 8.2.1j into 8.2.1i gives:

$$B = 0.2308 \frac{Ea^3}{\pi} f \qquad (8.2.1k)$$

Figure 8.2.3 shows the stresses computed from Eqs. 8.2.1e to 8.2.1h. The significance of these stresses is that generally, the different types of cracks observed in nominally brittle materials are a result of the action of different components of the stress field. For example, surface ring cracks are produced by σ_r radial tensile stresses ($\theta = \pi/2$); cracks that emanate from the corners of a pyramidal indentation are a result of the σ_ϕ hoop stress ($\theta = \pi/2$); median cracks beneath the indentation arise from outward σ_θ ($\theta = 0$) stresses along the axis of symmetry and lateral cracks from radial stresses σ_r ($\theta = 0$). Further details of these crack systems is given in Section 8.2.2.

Yoffe[6] presented a qualitative description of the residual stresses after unloading in which $\sigma_r = -0.42p_m$ and $\sigma_\phi = 0.12p_m$ on the surface, and $\sigma_r = 0.72p_m$ and of $\sigma_\phi = -0.06p_m$ on the axis beneath the indenter. Cook and Pharr[7] argue that upon unloading, the parameter B associated with the blister field is set at its maximum value corresponding to P_{max}. At $r = a_{max}$, the stresses now become dependent on the ratio P/P_{max} and the material properties f, E, and H. It is found that for small values of the product fE/H, radial cracking is predicted during unloading, but for large values, radial cracking may form during loading. Cook and Pharr[7] present details of the maximum stresses at $r = a_{max}$ for both loading and unloading for different values of fE/H. For the purposes of this section, we need only be aware of the complex nature of the stresses and their dependence on material parameters.

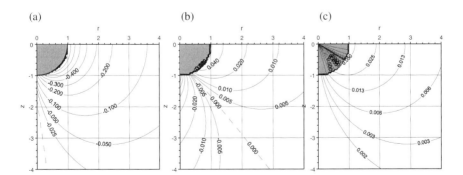

Fig. 8.2.3 Axis-symmetric stress distributions outside the plastic zone calculated using Eqs. 8.2.1e to 8.2.1h. Stresses have been normalized to mean contact pressure. Distances have been normalized to radius of circle of contact. (a) σ_r, (b) σ_θ, (c) σ_ϕ.

8.2.2 Indentation fracture

Fracture in brittle materials loaded with a pyramidal indenter occurs during both loading and unloading. Upon loading, tensile stresses are induced in the specimen material as the radius of the plastic zone increases. Upon unloading, additional stresses arise as the elastically strained material outside the plastic zone attempts to resume its original shape but is prevented from doing so by the permanent deformation associated with the plastic zone. However, the nature of the cracking depends on test conditions, and large variations in the number and location of the cracks that form in the specimen occur with only minor variations in the shape of the indenter, the loading rate, and the environment. There exists a large body of literature on the subject of indentation cracking with Vickers and other sharp indenters extending from the early 1970s to the late 1980s. A review of the field given by Cook and Pharr[7] in 1990 summarizes the details of the work done during this period.

For the present, it is sufficient for us to identify the types of cracks generally seen in specimens loaded with a Vickers or Berkovich indenter without being overly concerned about their initiation sequence (which depends on the experimental conditions). Generally, there are three types of crack, and they are illustrated in Fig. 8.2.4:

i. Radial cracks[†] are "vertical" half-penny type cracks[8] that occur on the surface of the specimen outside the plastic zone and at the corners of the residual impression at the indentation site. These radial cracks are formed by a hoop stress σ_ϕ ($\theta = \pi/2$) and extend downward into the specimen but are usually quite shallow (see Fig. 8.2.2 for coordinate system).

ii. Lateral cracks are "horizontal" cracks that occur beneath the surface and are symmetric with the load axis. They are produced by a tensile stress σ_r ($\theta = 0$) and often extend to the surface, resulting in a surface ring which may lead to chipping of the surface of the specimen.

iii. Median cracks are "vertical" circular penny cracks that form beneath the surface along the axis of symmetry and have a direction aligned with the corners of the residual impression. Depending on the loading conditions, median cracks may extend upward and join with surface radial cracks, thus forming two half-penny cracks which intersect the surface as shown in Fig. 8.2.4d. They arise due to the action of an outward stress σ_θ ($\theta = 0$).

As mentioned previously, the exact sequence of initiation of these three types of cracks is sensitive to experimental conditions. However, it is generally observed that in soda-lime glass (a popular and readily available test material)

[†] Shallow radial cracks are sometimes referred to as Palmqvist cracks after S. Palmqvist, who observed and described them in WC-Co specimens in 1957.

loaded with a Vickers indenter, median cracks initiate first. When the load is removed, the elastically strained material surrounding the median cracks cannot resume its former shape due to the presence of the permanently deformed plastic material (which leaves a residual impression in the surface of the specimen). Residual tensile stresses in the normal direction then produce a "horizontal" lateral crack which may or may not curve upward and intersect the specimen surface. Upon reloading, the lateral cracks close and the median cracks reopen. For low values of indenter load, radial cracks also form during unloading (in other materials, radial cracks may form during loading). For larger loads, upon unloading, the median cracks extend outward and upward and may join with the radial cracks to form a system of half-penny cracks, which are then referred to as "median/radial" cracks. In glass, the observed cracks at the corners of the residual impression on the specimen surface are usually fully formed median/radial cracks. However, in other brittle materials, with higher values of E/Y, radial cracks are usually quite distinct from the median cracks and form upon loading.

Experimental observations suggest that there is no one general sequence of indentation cracking. Chiang, Marshall, and Evans[4,5], and also Yoffe[6], have proposed analytical models that attempt to describe the cracking sequence in terms of a superposition of the Boussinesq elastic stress field and that associated with the expanding cavity model of Johnson. The Yoffe[6] model describes the evolution of stresses for both loading and unloading in terms of material and E/Y and appears to account for the appearance of radial and lateral cracking for loading and unloading sequences for a wide range of test materials. However, the model, as with that of Chiang, Marshall, and Evans[4,5], deals with axis-symmetric, pointed indenters. Clearly, the corners of a pyramidal indenter play an important role, since both median and radial cracks align themselves with the corners of the indentation. However, development of a three-dimensional analytical model that deals with this type of geometry would be an extremely difficult undertaking and would be best left to numerical finite-element analysis. The value of the axis-symmetric models is to provide physical insight, and that they do quite well.

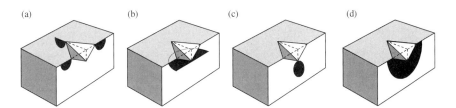

Fig. 8.2.4 Crack systems for Vickers indenter: (a) radial cracks, (b) lateral cracks, (c) median cracks, (d) half-penny cracks.

It is the radial and lateral cracks that are of particular importance, since their proximity to the surface has a significant influence on the fracture strength of the specimen. Fracture mechanics treatments of these types of cracks seek to provide a measure of fracture toughness based on the length of the radial surface cracks.

8.2.3 Fracture toughness

One of the main features of an indentation crack is that it is stable with increasing load[9]. While straight cracks in beam-type specimens are routinely used for fracture toughness testing of ductile materials, this type of test is very difficult to undertake with brittle materials. Attempts to do so usually end with catastrophic failure of the specimen. Further, indentation cracks require only a small surface test area, and multiple indentations can usually be made on the face of a single specimen. For these reasons, it is desirable to be able to measure fracture toughness using crack length data available from indentation tests.

Attention is usually given to the length of the radial cracks as measured from the corner of the indentation and then radially outward along the specimen surface as shown in Fig. 8.2.5.

Palmqvist[8] noted that the crack length "l" varied as a linear function of the indentation load. Lawn, Evans, and Marshall[10] formulated a different relationship, where they treated the fully formed median/radial crack and found the ratio $P/c^{3/2}$ (where c is measured from the center of contact to the end of the corner radial crack) is a constant, the value of which depends on the specimen material. Fracture toughness is found from:

$$K_c = k \left(\frac{E}{H}\right)^n \frac{P}{c^{3/2}} \qquad (8.2.3a)$$

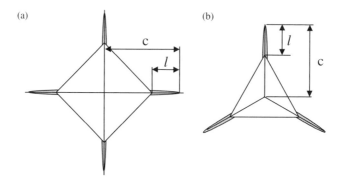

Fig. 8.2.5 Crack parameters for Vickers and Berkovich indenters. Crack length c is measured from the center of contact to end of crack at the specimen surface.

where k is a calibration constant equal to 0.016 and n = ½ with c = l+a. Various other studies have since been performed, and Anstis, Chantikul, Lawn, and Marshall[11] determined n = 3/2 and k = 0.0098. In 1987, Laugier[12-14] undertook an extensive review of previously reported experimental results and determined that:

$$K_c = x_v (a/l)^{1/2} \left(\frac{E}{H}\right)^{2/3} \frac{P}{c^{3/2}}. \qquad (8.2.3b)$$

With $x_v = 0.015$, Laugier showed that the radial and half-penny models make almost identical predictions of the dependence of crack length on load (note the similarity between Eqs. 8.2.3a and 8.2.3b. Experiments show that the term $(a/l)^{1/2}$ shows little variation between glasses (median/radial) and ceramics (radial). The significance of this result is that it is thus generally not possible to infer the existence of a fully formed median/radial crack from the observable crack length, and for opaque materials it is necessary to undertake sectioning of the specimen to obtain a full knowledge of the crack system in any particular material.

8.2.4 Berkovich indenter

The vast majority of toughness determinations using indentation techniques are performed with a Vickers diamond pyramid indenter. This indenter takes the form of a square pyramid with opposite faces at an angle of 136° (edges at 148°). However, the advantages of a Berkovich indenter have become increasingly important, especially in ultra-micro-indentation work, where the faces of the pyramid are more likely to meet at a single point rather than a line. However, despite this advantage, the loss of symmetry presents some problems in determining specimen toughness because half-penny cracks can no longer join two corners of the indentation. Ouchterlony[15] investigated the nature of the radial cracking emanating from a centrally loaded expansion star crack and determined a modification factor for stress intensity factor to account for the number of radial cracks formed.

$$k_1 = \sqrt{\frac{n/2}{1+\frac{n}{2\pi}\sin\frac{2\pi}{n}}}. \qquad (8.2.4a)$$

As proposed by Dukino and Swain[16], this modification has relevance to the crack pattern observed from indentations with a Berkovich indenter. The ratio of k_1 values for n = 4 (Vickers) and n=3 (Berkovich) is 1.073 and thus the length of a radial crack (as measured from the center of the indentation to the crack tip) from a Berkovich indenter should be $1.073^{2/3} = 1.05$ that from a Vickers indenter for the same value of K_1[16]. The Laugier expression can thus be written:

$$K_c = 1.073 x_v (a/l)^{1/2} \left(\frac{E}{H}\right)^{2/3} \frac{P}{c^{3/2}}. \qquad (8.2.4b)$$

8.3 Spherical Indenter

We turn now to a consideration of the elastic-plastic indentation stress field associated with a spherical indenter. As mentioned previously, analytical treatments of the indentation stress field for elastic-plastic contact are made complex by the presence of plasticity beneath the indenter. However, these difficulties can be circumvented by making use of the finite-element method[17-21]. Such modeling is of considerable importance because it provides numerical data for indentations involving complex geometries and material properties (see Chapter 12). As will be seen in Chapter 9, the actual shape and size of the plastic zone depend on the mechanical properties of the specimen material, particularly the ratio E/Y, where E is the elastic modulus and Y is the yield stress[22].

Of particular interest is a comparison between the elastic and elastic-plastic stress distributions shown in Fig. 8.3.1. This figure shows that the presence of the "plastic" or damage zone significantly alters the near-field indentation stresses. In general, the magnitudes of the maximum principal stresses appear to shift outward, away from the center of contact when compared to the elastic case. The far-field stresses, however, appear to be little changed from the elastic case. The shift in magnitudes of stresses away from the center of contact, indicates an outward shift in the distribution of upward pressure which serves to support the indenter. This is reflected in the contact pressure distributions shown in Fig. 8.3.2, where the contact pressure for the elastic-plastic case, appears to be more uniformly distributed (except for the rise in contact pressure near the edge of the contact circle) compared to the elastic case. In Fig. 8.3.2, results are plotted normalized to the mean contact pressure $p_m = 3.0$ GPa and contact radius $a_o = 0.326$ mm for the elastic case. Stresses have been calculated for a yield stress Y = 770 MPa with the load P = 1000N and R = 3.18mm. Absolute values may be obtained by multiplying by these factors.

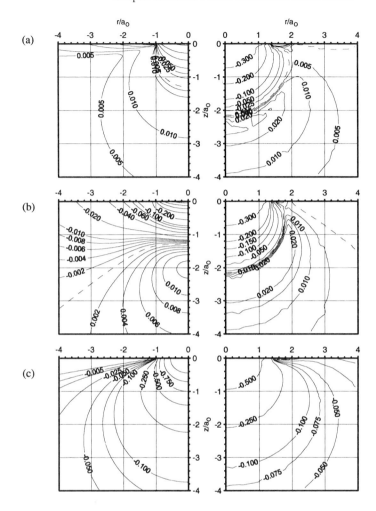

Fig. 8.3.1 Contours of principal stress for glass-ceramic material with spherical indenter. Left side of figure shows elastic solutions from Eqs. 5.4.2i to 5.4.2o in Chapter 5. Right side of figure shows finite-element results for elastic-plastic response. Magnitudes are shown for (a) σ_1, (b) σ_2, (c) σ_3. For both elastic and elastic-plastic results, distances are expressed relative to the contact radius $a_o = 0.326$ mm and stresses in terms of the mean contact pressure $p_m = 3.0$ GPa for the elastic case at P = 1000 N (after reference 22).

Figure 8.3.3 shows the variation in stress along the surface and downward along the axis of symmetry. As for Fig. 8.2.3, results are plotted normalized to the elastic contact pressure ($p_m = 3.0$ GPa) and contact radius ($a_o = 0.326$ mm). Also shown in this figure are the results for the elastic solution for the maximum principal stress σ_1. As shown in Fig. 8.3.3 (a), along the surface, the maximum value of σ_1 is very much the same as for the elastic case, although the contact pressure is correspondingly lower due to the larger area of contact in the elastic-plastic case (with a = 0.437 mm). There is an interesting change in magnitude for all stresses within the contact zone near the edge of the circle of contact. Later we shall see that this arises due to the plastic zone being contained within the area of the circle of contact, and this discontinuity flattens out for contacts on materials, such as metals, where the plastic zone extends beyond the radius of circle of contact.

In Fig. 8.3.3 (b), the variation in stresses along the axis of symmetry downward into the specimen material is shown. Note that the maximum tensile stress occurs at the elastic-plastic boundary and is larger by a factor of ≈3.6 than that calculated for an elastic contact.

The significance of these results is particularly important when we wish to use indentation stress fields for structural reliability analysis. The failure of brittle materials is often attributed to the action of tensile stresses on surface flaws. For example, the procedure for determining the conditions for initiation of a Hertzian crack in soda-lime glass (presented in Chapter 7) applies to a purely elastic contact. For the type of material described here—a nominally brittle ceramic which undergoes shear-driven failure within a "plastic" zone—an analysis involving Weibull statistics may not be appropriate or may need to be applied to flaws *within* the interior of the specimen rather than on the surface. Whatever analysis is selected, a detailed knowledge of the state of stress within the material is required.

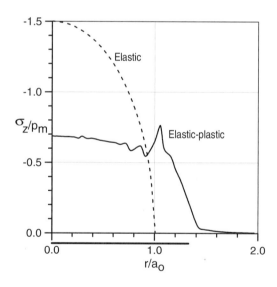

Fig. 8.3.2 Contact pressure distribution for elastic (equation), and elastic-plastic (finite-element) contact for P = 1000 N, R = 3.18 mm. Results are normalized to a_o and p_m as in Fig. 8.3.1. The bar at the bottom of the horizontal axis indicates the radius of circle of contact a = 0.437 mm for the elastic-plastic condition (after reference 22).

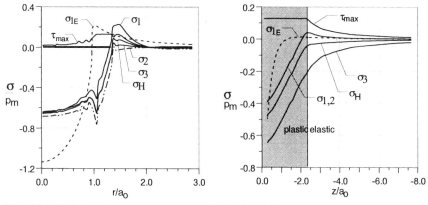

Fig. 8.3.3 Variation in stresses σ_1, σ_2, σ_3, the hydrostatic component σ_H, and maximum shear stress τ_{max} along (a) the surface of the specimen at $z = 0$ and (b) downward along the axis of symmetry at $r = 0$. In both (a) and (b), the dashed curve σ_{1E} shows the variation in σ_1 as calculated from elastic formulas for comparison with the elastic-plastic finite-element result. The horizontal bar in (a) indicates the radius of circle of contact $a = 0.437$ mm for the elastic-plastic condition. The shaded area in (b) indicates the region in which plastic strains have occurred (after reference 22).

References

1. D.M. Marsh, "Plastic flow in glass," Proc. R. Soc. London, Ser. A279, 1964, pp. 420–435.
2. R. Hill, *The Mathematical Theory of Plasticity*, Clarendon Press, Oxford, 1950.
3. K.L. Johnson, "The correlation of indentation experiments," J. Mech. Phys. Solids 18, 1970, pp. 115–126.
4. S.S. Chiang, D.B. Marshall, and A.G. Evans, "The response of solids to elastic/plastic indentation. 1. Stresses and residual stresses," J. Appl. Phys. 53 1, 1982, pp. 298–311.
5. S.S. Chiang, D.B. Marshall, and A.G. Evans, "The response of solids to elastic/plastic indentation. 2. Fracture initiation," J. Appl. Phys. 53 1, 1982, pp. 312–317.
6. E.H. Yoffe, "Elastic stress fields caused by indenting brittle materials," Philos. Mag. A, 46, 1982, pp. 617–628.
7. R.F. Cook and G.M.Pharr, "Direct observation and analysis of indentation cracking in glasses and ceramics," J. Am. Ceram. Soc. 73 4, 1990, pp. 787–817.
8. S. Palmqvist, "A method to determine the toughness of brittle materials, especially hard materials," Jernkontorets Ann. 141, 1957, pp. 303–307.
9. F.C. Roesler, "Brittle fractures near equilbrium," Proc. R. Soc. London, Ser. B69 1956, pp. 981–992.
10. B.R. Lawn, A.G. Evans, and D.B. Marshall, "Elastic/plastic indentation damage in ceramics: the median/radial crack system," J. Am. Ceram. Soc. 63, 1980, pp. 574–581.

11. G.R. Anstis, P.Chantikul, B.R.Lawn, and D.B.Marshall, "A critical evaluation of indentation techniques for measuring fracture toughness: I Direct crack measurements," J. Am. Ceram. Soc. 64 9, 1981, pp. 533–538.
12. M.T. Laugier, "Palmqvist indentation toughness in WC-Co composites," J. Mater. Sci. Lett. 6, 1987, pp. 897–900.
13. M.T.Laugier, "Palmqvist toughness in WC-Co composites viewed as a ductile/brittle transition," J. Mater. Sci. Lett. 6, 1987, pp. 768–770.
14. M.T.Laugier, "New formula for indentation toughness in ceramics," J. Mater. Sci. Lett. 6, 1987, pp. 355–356.
15. F. Ouchterlony, "Stress intensity factors for the expansion loaded star crack," Eng. Frac. Mechs. 8, 1976, pp. 447–448.
16. R. Dukino and M.V.Swain, "Comparative measurement of indentation fracture toughness with Berkovich and Vickers indenters," J. Am. Ceram. Soc. 75 12, 1992, pp. 3299–3304.
17. C. Hardy, C.N. Baronet, and G.V. Tordion, "The elastic-plastic indentation of a half-space by a rigid sphere," Int. J. Numer. Methods Eng. 3, 1971, pp. 451–462.
18. P.S. Follansbee and G.B. Sinclair, "Quasi-static normal indentation of an elasto-plastic half-space by a rigid sphere-I," Int. J. Solids Struct. 20, 1981, pp. 81–91.
19. R. Hill, B. Storakers and A.B. Zdunek, "A theoretical study of the Brinell hardness test," Proc. R. Soc. London, Ser. A423, 1989, pp. 301–330.
20. K. Komvopoulos, "Finite element analysis of a layered elastic solid in normal contact with a rigid surface," J. Tribology, Trans. ASME, 111, 1988, pp. 477–485.
21. K. Komvopoulos, "Elastic-plastic finite element analysis of indented layered media," J. Tribology, Trans. ASME, 111, 1989, pp. 430–439.
22. G. Caré and A.C.Fischer-Cripps, "Elastic-plastic indentation stress fields using the finite element method," J. Mater. Sci. 32, 1997, pp. 5653–5659.

Chapter 9
Hardness

9.1 Introduction

Indentation tests involving hard, spherical indenters have been the basis of hardness testing since the time of Hertz in 1881[1,2]. Conventional indentation hardness tests involve the measurement of the size of a residual plastic impression in the specimen as a function of the indenter load. Theoretical approaches to hardness can generally be categorized according to the properties of the indenter and the assumed response of the specimen material. For sharp indenters, the specimen is usually approximated by a rigid-plastic material in which plastic flow is assumed to be governed by flow velocity considerations. For blunt indenters, the specimen responds in an elastic-plastic manner, and plastic flow is usually described in terms of the elastic constraint offered by the surrounding material. The hardness of brittle materials is conveniently measured using a diamond pyramid indenter, since a residual impression is readily obtained at relatively low values of indenter load. The use of spherical indenters in hardness measurements is usually restricted to tests involving ductile materials but has recently been applied to certain brittle ceramics that exhibit a shear-driven "plasticity" in the indentation stress field.

9.2 Indentation Hardness Measurements

9.2.1 Brinell hardness number

The Brinell test involves the indentation of the specimen with a hard spherical indenter. The method was developed by J.A. Brinell in 1900[3]. For an indenter diameter D, and diameter of the residual impression d, the hardness number is calculated from:

$$\text{BHN} = \frac{2P}{\pi D \left(D - \sqrt{D^2 - d^2} \right)} \qquad (9.2.1a)$$

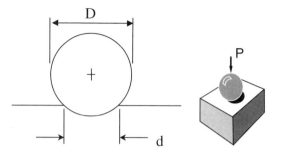

Fig. 9.2.1 Geometry of a Brinell hardness test.

This formula gives the hardness as a function of the load and the actual surface area of the residual impression. The diameter of the impression d is measured in the plane of the original surface. Brinell hardness measurements are usually made such that the ratio d/D is between 0.25 and 0.5 and the load applied for at least 30 seconds. For iron and steel, a ball diameter of 10 mm with a load of 30kN is commonly used. Figure 9.2.1 shows the relevant features of the test geometry.

The Brinell hardness number is favored by some engineers because of the existence of an empirical relationship between it and the ultimate tensile strength of the specimen material[*]. However, for physical reasons, it is preferable to use the projected contact area rather than the actual surface area of the specimen as the divisor since this gives the pressure beneath the indenter which opposes the applied force.

9.2.2 Meyer hardness

The Meyer hardness is similar to the Brinell hardness except that the projected area of contact rather than the actual curved surface area is used to determine the hardness. In this case, the hardness number is equivalent to the mean contact pressure between the indenter and the surface of the specimen. As we shall see, the mean contact pressure is a quantity of considerable physical significance. The Meyer hardness is given by:

$$H = p_m = \frac{4P}{\pi d^2} \qquad (9.2.2a)$$

where d is the diameter of the contact circle at full load (assumed to be equal to the diameter of the residual impression in the surface—see Chapter 10).

---

[*] Ultimate tensile strength is defined as the maximum nominal stress measured using the original cross-sectional area of the test specimen.

9.2.3 Vickers diamond hardness

The Vickers diamond indenter takes the form of a square pyramid with opposite faces at an angle of 136° (edges at 148°). The indenter was suggested by Smith and Sandland in 1924[4]. The Vickers diamond hardness, VDH, is calculated using the indenter load and the actual surface area of the impression. The area of the base of the pyramid, at a plane in line with the surface of the specimen, is equal to 0.927 times the surface area of the faces that contact the specimen. The mean contact pressure p_m is given by the load divided by the projected area of the impression[†]. Thus, the Vickers hardness number is lower than the mean contact pressure by ≈ 7%. The Vickers diamond hardness is found from:

$$\text{VDH} = \frac{2P}{d^2} \sin \frac{136°}{2} = 1.86 \frac{P}{d^2} \tag{9.2.3a}$$

with d equal to the length of the diagonal measured from corner to corner as shown in Fig. 9.2.3.

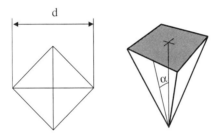

Fig. 9.2.3 Geometry of Vickers hardness test.

9.2.4 Knoop hardness

The Knoop indenter is similar to the Vickers indenter except that the diamond pyramid has unequal length edges, resulting in an impression that has one diagonal with a length seven times the shorter diagonal. The angles for the opposite faces of a Knoop indenter are 172°30′ and 130°. The Knoop indenter is particularly useful for the study of highly brittle materials due to the smaller depth of penetration for a given indenter load. Further, due to the unequal lengths of the diagonals, it is also very useful for investigating anisotropy of the surface of the specimen. The indenter was invented in 1934 at the National Bureau of Standards in the United States by F. Knoop, C.G. Peters, and W.B. Emerson[5].

[†] Assuming frictionless contact between the indenter and the specimen.

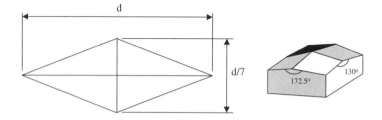

Fig. 9.2.4 Geometry of a Knoop indenter.

As shown in Fig. 9.2.4, the length d of the longer diagonal is used to determine the projected area of the impression. The Knoop hardness number calculated from:

$$\text{KHN} = \frac{2P}{d^2 \left[\cot \dfrac{172.5}{2} \tan \dfrac{130}{2} \right]} \qquad (9.2.4a)$$

9.2.5 Other hardness test methods

The Rockwell hardness test was introduced in 1920, and the hardness value is obtained from the depth of the indentation of an indenter of specified geometry. The major advantage of this type of test is its simplicity, since the hardness is read as a number from a dial gauge on the instrument. The Rockwell hardness tester measures the depth of the permanent impression left by the indenter, which is usually either a steel ball of specified diameter or a diamond tipped cone of specified radius of curvature.

Another popular instrument is the Shore scleroscope. This instrument, devised by A.F. Shore in 1906, allows the indenter to fall onto the test specimen, and the hardness is given in terms of the rebound height of the indenter. The height through which the indenter falls is held constant.

9.3 Meaning of Hardness

The meaning of hardness has been the subject of considerable attention by scientists and engineers since the early 1700s. It was appreciated very early on that hardness indicated a resistance to penetration or permanent deformation. Early methods of measuring hardness, such as the scratch method, although convenient and simple, were found to involve too many variables to provide the means for a scientific definition of hardness. Static indentation tests involving spherical or conical indenters were first used as the basis for theories of hardness. Compared to "dynamic" tests, static tests enabled various criteria of hardness to be established since the number of test variables was reduced to a manageable level. The most famous criterion is that of Hertz[2], who postulated that an absolute value for hardness was the least value of pressure beneath a spherical indenter necessary to produce a permanent set at the center of the area of contact. Later treatments by Auerbach[6], Meyer[7], and Hoyt[8] were all directed to removing some of the practical difficulties in Hertz's original proposal. As we shall see, more than 100 years later, an absolute definition of hardness is still open to investigation.

9.3.1 Compressive modes of failure

Indentation hardness measurements are concerned with the residual impression in the surface of the test specimen. Such impressions arise from the plastic deformation of material in shear in the compressive stress field beneath the indenter and are generally easily obtained with ductile materials. In uniaxial tests, compressive failure with ductile materials is at once recognized by "barreling" and failure is due to yielding, which can be characterized by the compressive yield strength of the material. Generally, the compressive yield strength for ductile materials does not change significantly with varying loading mechanisms and can be regarded as a material property.

For brittle specimens subjected to compressive loading, three modes of failure have been identified: axial splitting[9], where failure initiates from a fissure oriented in the same direction as the compressive stress; shear fracture, first studied in detail by Coulomb[10], where failure occurs at an angle to the applied compressive stress; and ductile failure[11], which can occur in some brittle materials in the presence of a confining pressure. These modes of failure are illustrated in Fig. 9.3.1. It is ductile failure that is of interest in indentation hardness testing of brittle materials.

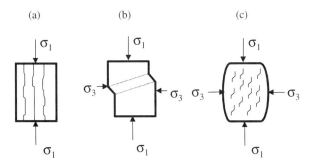

Fig. 9.3.1 Compressive modes of failure. (a) Slabbing occurs in the absence of confining pressure ($\sigma_3 = 0$). (b) At moderate confining pressure, cracks nucleate within the material to form a "shear zone" and (c) at high confining pressures, distributed microcracking is in evidence[12].

Experiments[12,13] show that the mode of compressive failure in brittle materials depends on the degree of confining pressure. With no confining pressure, a brittle specimen loaded in uniaxial compression usually fails by axial splitting. With a small confining pressure, failure usually occurs as a single shear fracture oriented at $\approx 45°$ to the line of action of the compressive stress. When a confining pressure of a magnitude comparable with the applied compressive stress is applied, failure occurs as a series of multiple shear fractures distributed throughout the specimen volume. The material is said to be "ductile" due to the similarity of shape of the uniaxial stress-strain curve with that obtained on tests with ductile materials.

9.3.2 The constraint factor

Static indentation hardness tests usually involve the application of load to a spherical or diamond pyramid indenter. The pressure distribution beneath the indenter is of particular interest. The value of the mean contact pressure p_m at which there is no increase with increasing indenter load is related to the hardness number H. For hardness methods that employ the projected contact area, the hardness number H is given directly by the mean pressure p_m. Experiments show that for the hardness methods mentioned in Section 9.3, the mean pressure between the indenter and the specimen is directly proportional to the material's yield, or flow stress in compression and can be expressed as:

$$H \approx CY \qquad (9.3.2a)$$

where Y is the yield, or flow stress, of the material‡. The mean contact pressure in an indentation test is higher than that required to initiate yield in a uniaxial compression test since it is the shear component of stress that is responsible for plastic flow. The maximum shear stress is equal to half the difference between the maximum and minimum principal stresses, and in an indentation stress field, where the stress material is constrained by the surrounding matrix, there is a considerable hydrostatic component. Thus, the mean contact pressure is greater than that required to initiate yield compared to a uniaxial compressive stress. It is for this reason that C in Eq. 9.3.2a is called the "constraint factor," the value of which depends upon the type of specimen, the type of indenter and other experimental parameters. For the indentation methods mentioned here, both experiments and theory predict $C \approx 3$ for materials with a large value of the ratio E/Y (e.g., metals)[14]. For low values of E/Y (e.g., glasses[15,16]), $C \approx 1.5$. The flow, or yield stress Y, in this context is the stress at which plastic yielding first occurs.

---

‡ The terms yield stress and flow stress are used interchangeably in this book.

9.3.3 Indentation response of materials

The nature of the constraint factor, which relates hardness to yield stress, has been the subject of considerable scientific research. The aim of such research has been to explain the origin and value for the constraint factor in terms of the indenter geometry and mechanical and physical properties of the test specimens. As we have seen, hardness is intimately related to the mean contact pressure p_m beneath the indenter. In general, the mean contact pressure depends on the load and the geometry of the indenter. Valuable information about the elastic and plastic properties of a material can be obtained with spherical indenters when the mean contact pressure, called the "indentation stress," is plotted against the ratio a/R, where a is the contact area radius and R is the indenter radius. The quantity a/R is termed the "indentation strain." The indentation stress-strain response of an elastic-plastic solid can generally be divided into three regimes which depend on the uniaxial compressive yield stress Y of the material.

1. $p_m < 1.1Y$—full elastic response, no permanent or residual impression left in the test specimen after removal of load.
2. $1.1Y < p_m < CY$—plastic deformation exists beneath the surface but is constrained by the surrounding elastic material, where C is a constant whose value depends on the material and the indenter geometry.
3. $p_m = CY$—plastic region extends to the surface of the specimen and continues to grow in size such that the indentation contact area increases at a rate that gives little or no increase in the mean contact pressure for further increases in indenter load.

In Region 1, during the initial application of load, the response is elastic and can be predicted from the Hertz relation[1]:

$$p_m = \left(\frac{3E}{4\pi k}\right)\frac{a}{R} \tag{9.3.3a}$$

In Eq. 9.3.3a, k is a dimensionless constant given by:

$$k = 9/16\left[\left(1-v^2\right)+\left(1-v'^2\right)E/E'\right] \tag{9.3.3b}$$

where v and v' are Poisson's ratio, and E and E' are Young's modulus of the specimen and the indenter, respectively. Equation 9.3.3a assumes a condition of linear elasticity and makes no allowance for yield within the specimen material. For a fully elastic response, the principal shear stress for indentation with a spherical indenter is a maximum at $\approx 0.47 p_m$ at a depth of $\approx 0.5a$ beneath the specimen surface directly beneath the indenter[17]. Following Tabor[14], we may employ either the Tresca or von Mises shear stress criteria, where plastic flow occurs at $\tau \approx 0.5Y$, to show that plastic deformation in the specimen beneath a spherical indenter can be first expected to occur when $p_m \approx 1.1Y$. Figure 9.3.2 shows the surface and section view of the results of an indentation test using a spherical indenter at very low load on a specimen of micaceous glass-ceramic— a nominally brittle material that exhibits shear-driven plasticity in the indentation stress field. Note that shear-driven damage occurs initially beneath the specimen surface.

Theoretical treatment of events within Region 2 is difficult because of the uncertainty regarding the size and shape of the evolving plastic zone. At high values of indentation strain (Region 3), the mode of deformation appears to depend on the type of indenter and the specimen material. The presence of the free surface has an appreciable effect, and the plastic deformation within the specimen is such that, assuming no work hardening, little or no increase in p_m occurs with increasing indenter load.

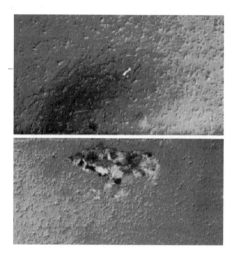

Fig. 9.3.2 Indentation test result for glass-ceramic material, E/Y = 90, at low load. Top is the surface view showing residual impression, and bottom is a section view showing subsurface shear-driven damage occurring beneath the specimen surface. Indenter load of P = 100 N and indenter radius R = 3.18 mm.

9.3.4 Hardness theories

Theoretical approaches to hardness can generally be categorized according to the characteristics of the indenter and the assumed response of the specimen material. For sharp indenters, the specimen is usually approximated by a rigid-plastic material in which plastic flow is assumed to be governed by flow velocity considerations. For blunt indenters, the specimen responds in an elastic-plastic manner, and plastic flow is usually described in terms of the elastic constraint offered by the surrounding material. Various semiempirical models that describe experimentally observed phenomena at values of indentation strain at or near a condition of full plasticity have been given considerable attention in the literature[14,16,18-30]. These models variously describe the response of the specimen material in terms of slip lines, elastic displacements, and radial compressions. For sharp wedge or conical indenters, substantial upward flow is usually observed, and since elastic strains are thus negligible compared to plastic strains, the specimen can be regarded as being rigid plastic. A cutting mechanism is involved, and new surfaces are formed beneath the indenter as the volume displaced by the indenter is accommodated by the upward flow of plastically deformed material. The constraint factor C in this case arises due to flow and velocity considerations[18]. With blunt indenters, Samuels and Mulhearn[21] noted that the mode of plastic deformation at a condition of full plasticity appears to be a result of compression rather than cutting, and the displaced volume is assumed to be taken up entirely by elastic strains within the specimen material. This idea was given further attention by Marsh[20], who compared the plastic deformation in the vicinity of the indenter to that which occurs during the radial expansion of a spherical cavity subjected to internal pressure, an analysis of which was given previously by Hill[19]. The most widely accepted treatment is that of Johnson[23,30], who replaced the expansion of the cavity with that of an incompressible hemispherical core of material subjected to an internal pressure. Here, the core pressure is directly related to the mean contact pressure. This is the so-called "expanding cavity" model, which is worthy of or special attention.

9.3.4.1 Expanding cavity model

Johnson postulated that the analysis for the expansion of a spherical cavity in an elastic-plastic material could be applied to the hemispherical radial mode of deformation observed in indentation tests by replacing the cavity with an incompressible, hemispherical core of material directly beneath the indenter of radius equal to the contact circle. Surrounding the core is a hemispherical plastic zone, which connects with the elastically strained material at some larger radius c. As shown in Fig. 9.3.3, penetration of the indenter causes material particles at the outer boundary of the core, r = a, to be displaced radially outward an amount du(a) so that the volume swept out by the movement of these particles is equal to the volume of material displaced by the indenter. As the indenter penetrates the specimen, the radius of the core, assumed to be equal to the radius of the circle of contact, increases an amount da, which is greater than du(a). Material particles formerly at the outer boundary of the core are now in the interior, the core boundary having moved outward beyond them. But, it is important to note that it is the *radial displacements of particles at the initial core boundary* and not the movement of the boundary that gives rise to the volumetric compatibility with the movement of the indenter. The elastic-plastic boundary, at radius c, also extends an amount dc, which depends on the type of indenter. For geometrically similar indentations, such as a conical indenter, the ratio c/a is a constant for all values of load and penetration depths.

The volume of the material displaced by the indenter is accommodated by radial displacements du(a) of material at the moving boundary of the rigid, hydrostatic core. These displacements give rise to stresses that are sufficiently high to cause plastic deformation in the surrounding material—the plastic zone. Within the plastic zone, a < r < c, the stresses are given by Hill[19] as:

$$\frac{\sigma_r}{Y} = -2\ln\left(\frac{c}{r}\right) - \frac{2}{3}$$
$$\frac{\sigma_\theta}{Y} = -2\ln\left(\frac{c}{r}\right) + \frac{1}{3}$$
(9.3.4.1a)

The yield criterion in this case is simply $\sigma_r - \sigma_\theta = Y$. At the core boundary, r = a, the radial stress given by Eq. 9.3.4.1a is equal to the pressure p within the core (with a change in sign) so that:

$$\frac{\bar{p}}{Y} = \frac{2}{3} + 2\ln(c/a)$$
(9.3.4.1b)

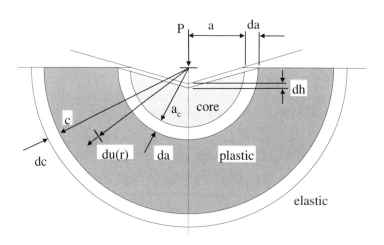

Fig. 9.3.3 Expanding cavity model schematic. The contacting surface of the indenter is encased by a hydrostatic "core" of radius a_c which is in turn surrounded by a hemispherical plastic zone of radius c. An increment of penetration dh of the indenter, results in an expansion of the core da and the volume displaced by the indenter is accommodated by radial movement of particles du(r) at the core boundary. This in turn causes the plastic zone to increase in radius by an amount dc (after reference 31).

Note that at $c = a$, Eq. 9.3.4.1b predicts yielding to occur first when $p = 2/3Y$. Outside the plastic zone, $r \geq c$,

$$\frac{\sigma_r}{Y} = -\frac{2}{3}\left(\frac{c}{r}\right)^3$$
$$\frac{\sigma_\theta}{Y} = \frac{1}{3}\left(\frac{c}{r}\right)^3$$
(9.3.4.1c)

Thus, within the plastic zone, the radial stress decreases in magnitude from that within the core, p, to $-2/3Y$ and the hoop stress correspondingly increases to satisfy the yield criterion to reach a maximum value of $1/3Y$ at the elastic-plastic boundary. Outside the plastic zone, both stresses decrease with increasing r, and the yield criterion is no longer met.

Radial displacements of material at the boundary of the core $r = a$, according to Hill[19], can be found from:

$$\frac{du(a)}{dc} = \frac{Y}{E}\left[3(1-v)\frac{c^2}{a^2} - 2(1-2v)\frac{a}{c}\right]$$
(9.3.4.1d)

Now, for an increment of penetration dh, the volume displaced by the indenter is equivalent to that of the movement of material particles on the boundary of the hydrostatic core, hence, for a conical indenter:

$$2\pi a^2 \, du(a) = \pi a^2 \, dh$$
$$= \pi a^2 \tan\beta \, da$$
(9.3.4.1e)

where β is the angle of inclination of the indenter with the specimen surface. Since:

$$\frac{du(a)}{dc} = \frac{du(a)}{da}\frac{da}{dc}$$
(9.3.4.1f)

we have:

$$\frac{Y}{E}\left[3(1-v)\frac{c^2}{a^2} - 2(1-2v)\frac{a}{c}\right] = \frac{1}{2}\tan\beta\frac{da}{dc}$$
(9.3.4.1g)

For geometrically similar indentations, such as with a conical indenter, the radius of the plastic zone increases at the same rate as that of the core, hence, $da/dc = a/c$. Using this result in Eq. 9.3.4.1g and substituting the resulting expression for c/a into Eq. 9.3.4.1b we obtain:

$$\frac{\bar{p}}{Y} = \frac{2}{3}\left[1 + \ln\left(\frac{(E/Y)\tan\beta + 4(1-2v)}{6(1-v^2)}\right)\right]$$
(9.3.4.1h)

where p is the pressure within the core and is related to the mean contact pressure beneath the indenter. For the case of a spherical indenter, Johnson[30] suggests that $\tan\beta$ in Eq. 9.3.4.1h can be replaced with a/R for small values of β, since Eq. 9.3.4.1e applies equally well to both conical and spherical indenters (where the depth of penetration for a cone is twice that of a sphere for the same value of a—see Section 9.5.2). However, Eq. 9.3.4.1h with $\tan\beta$ = a/R is not appropriate since it was formulated using the geometrical similarity of the stress field associated with a conical indenter (i.e., da/dc = c/a). This is not true for the case of the spherical indenter, and hence we need to consider the more general equation:

$$\frac{\bar{p}}{Y} = \frac{2}{3}\left[1 + \ln\left(\frac{(E/Y)\left[\frac{da}{dc}\frac{c}{a}\right]\frac{a}{R} + 4(1-2v)}{6(1-v^2)}\right)\right] \qquad (9.3.4.1i)$$

We require information concerning the product of da/dc and c/a. We may expect that since the elastic stress distribution within the specimen for a spherical indenter is directly proportional to a, then if c/a = Ka, where K is a constant, then dc/da = 2Ka and hence:

$$\frac{\bar{p}}{Y} = \frac{2}{3}\left[1 + \ln\left(\frac{(E/Y)\frac{1}{2}\frac{a}{R} + 4(1-2v)}{6(1-v^2)}\right)\right] \qquad (9.3.4.1j)$$

Equation 9.3.4.1j relates the core pressure p and the ratio a/R for a spherical indenter based on the assumption that c/a = Ka.

As noted previously, in the case of a spherical indenter, the transition between elastic and full plastic response occurs as a result of yielding of elastically constrained material some distance beneath the surface of the specimen at some finite value of contact radius a*. Swain and Hagan[16] suggested therefore that $\tan\beta$ in Eq. 9.3.4.1h should be replaced by (a–a*)/R but doing so not only ignores the condition of nongeometrical similarity associated with a spherical indenter but also violates the volumetric compatibility specified by Johnson[23]. If appropriate adjustments are made to Eq. 9.3.4.1h to account for both the geometry of the indentation and the finite value of the contact radius at the initiation of yield, we obtain:

$$\frac{\bar{p}}{Y} = \frac{2}{3}\left[1 + \ln\left(\left((E/Y)\frac{1}{2}\frac{a}{a'}\left[\frac{a^2}{a'^2} - \frac{a^{*2}}{a'^2}\right]\frac{a}{R} + 4(1-2v)\right)\bigg/6(1-v^2)\right)\right]$$

$$(9.3.4.1k)$$

where a' = a–a* and is the effective radius of the core.

The core pressure is directly related to the mean contact pressure beneath the indenter and according to Johnson[30] is given by:

$$p_m = \bar{p} + \frac{2}{3}Y \qquad (9.3.4.1l)$$

The size of the plastic zone c/a can be found from Eq. 9.3.4.1g. We should note in passing that the expanding cavity model requires the distribution of pressure across the face if the indenter is uniform and equal to p_m.

9.3.4.2 The elastic constraint factor

An alternative to the expanding cavity model is given by Shaw and DeSalvo[24,25], who showed that the observed region of plasticity in their bonded-interface specimens was evidence of an elastically constrained mode of deformation. Like the expanding cavity model, the specimen material is assumed to behave in an elastic-plastic manner, and the volume displaced by the indenter is ultimately taken up by elastic displacements in the specimen material remote from the indentation. By comparing the elastic stress field for a spherical indenter and that for an equivalent inverted wedge, Shaw and DeSalvo argue that the perimeter of the fully developed plastic zone is restricted to passing through the edge of the contact circle at the specimen surface. Figure 9.3.4 shows the results of an indentation experiment, using a bonded-interface, or split-specimen, technique on a mica-containing glass-ceramic showing the correspondence between the shape of the plastic zone and the elastic stress field. The shape of the zone is similar to that observed by Shaw and DeSalvo with a metal specimen.

However, Shaw and DeSalvo do not present quantitative data in the form of an indentation stress-strain curve, nor do they offer an analytical expression of such a relationship equivalent to Eq. 9.3.4.1k. Rather, they present a method for determining the constraint factor C which they imply is independent of the indention strain—the only proviso being that the plastic zone be fully developed. It can be seen from Fig. 9.3.4 that the edge of the plastic zone corresponds to the elastic stress contour such that $\tau_{max}/p_m = 0.23$ and that plasticity occurs when $\tau_{max} = Y/2$, where Y is the yield stress of the specimen material. Now, since $p_m = CY$ for a condition of full plasticity, then:

$$\frac{Y}{2}\frac{1}{p_m} = 0.23 \therefore$$
$$C = 2.2 \qquad (9.3.4.2a)$$

The theory appears to be inconsistent with their requirement that the pressure distribution across the contact area be unchanged from the Hertzian, or fully elastic, case which predicts p_m directly proportional to a/R.

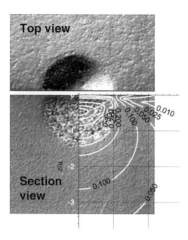

Fig. 9.3.4 Elastic constraint theory demonstrated for glass-ceramic material. Contours of normalized maximum shear stress calculated using Hertzian elastic stress field have been overlaid onto the section view of subsurface damage beneath the indentation. Elastic-plastic boundary appears to coincide with $\tau_{max}/p_m \approx 0.23$, leading to a constraint factor C ≈ 2.2 for this material.

9.3.4.3 Region 3: Rigid-plastic—Slip line theory

When the free surface of the specimen begins appreciably to influence the shape of the plastic zone, and the plastic material is no longer elastically constrained, the volume of material displaced by the indenter is accommodated by upward flow around the indenter. The specimen then takes on the characteristics of a rigid-plastic solid, since any elastic strains present are very much smaller than the plastic flow of unconstrained material. Plastic yield within such a material depends upon a critical shear stress which may be calculated using either of the von Mises or Tresca failure criteria. In the slip-line field solution, developed originally in two dimensions by Hill, Lee, and Tupper[18], the volume of material displaced by the indenter is accounted for by upward flow, as shown in Fig. 9.3.5. This upward flow requires relative movement between the indenter and the material on the specimen surface, and hence the solution depends on friction at this interface.

Figure 9.3.5 shows the situation for frictionless contact. The material in the region ABCDE flows upward and outward as the indenter moves downward under load. Since frictionless contact is assumed, the direction of stress along the line AB is normal to the face of the indenter. The lines within the region ABDEC are oriented at 45° to AB and are called "slip lines" (lines of maximum shear stress). Hill, Lee, and Tupper[18] formulated a mathematical treatment of the two-dimensional case of Fig. 9.3.5. If the indenter is assumed to be penetrating the specimen with a constant velocity, and if geometrical similarity is maintained (see Chapter 10), the angle ψ can be chosen so that the velocities of elements of material on the free surface, contact surface, and boundary of the rigid plastic material are consistent. Note that this type of indentation involves a "cutting" of the specimen material along the line 0A and the creation of new surfaces which travel upward along the contact surface. The contact pressure across the face of the indenter[§] is given by:

$$p_m = 2\tau_{max}(1+\alpha)$$
$$= H$$
(9.3.4.3a)

where τ_{max} is the maximum value of shear stress in the specimen material and α is the cone semi-angle (in radians). Invoking the Tresca shear stress criterion, where plastic flow occurs at $\tau_{max} = 0.5Y$, and substituting into Eq. 9.6, gives:

$$H = Y(1+\alpha)$$
$$\therefore$$
$$C = (1+\alpha)$$
(9.3.4.3b)

We refer to the constraint factor determined by this method as C_{flow} and as such it is a "flow" constraint. For values of α between 70° and 90°, Eq. 9.3.4.3b gives only a small variation in C_{flow} of 2.22 to 2.6. Friction between the indenter and the specimen increases the value of C_{flow}. A slightly larger value for C_{flow} is found when the von Mises stress criterion is used (where $\tau_{max} \approx 0.58Y$). For example, at $\alpha = 90°$, Eq. 9.3.4.3b with the von Mises criterion gives $C = 3$.

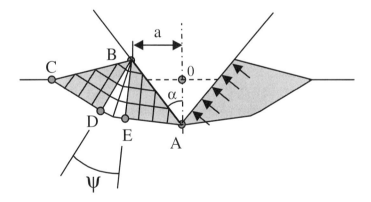

Fig. 9.3.5 Slip-line theory.

[§] Tabor shows that for a fully plastic state in three dimensions, the pressure distribution across the face of a cylindrical indenter is not uniform but higher at the center of the contact area.

Experiments[21] show that the shape of the plastic deformation in metals, near a pyramidal indenter, does not follow that predicted by the theory when the cone semiangle is greater than about 60–70°. In particular, as the indenter becomes less sharp (i.e., larger cone semi-angles), the displacement of material upward in the vicinity of the indenter is significantly less than that predicted by the theory. Hardness testing is usually performed with indenters with a cone semi-angle greater than 60°, and the failure of the slip-line theory to account for the observed deformations somewhat downgrades the applicability of the theory under these conditions.

9.3.4.4 Region 3: Elastic-brittle—Compaction and densification

Plastic deformation is normally associated with ductile materials that behave in an elastic-plastic fashion. Brittle materials generally exhibit elastic behavior, and fracture occurs rather than plastic yielding at high loads. However, plastic deformation is routinely observed in brittle materials, such as glass, beneath the point of a diamond pyramid indenter. The mode of plastic deformation is considerably different from that occurring in metals. In brittle materials, plastic deformation is more likely to be a result of densification, where the specimen material undergoes a phase change as a result of the high value of compressive stress beneath the indenter[16]. The Tabor relationship, which relates yield stress to hardness, with $C \approx 3$ applies to metals, where plastic flow occurs as a result of slippage of crystal planes and dislocation movement, and may not be so appropriate for determining the yield strength of brittle solids.

9.3.4.5 Comparison of the models

It is generally accepted that the mode of deformation experienced by specimens in an indentation hardness test depends on the characteristics of the indenter and the specimen material. Indenters whose tangents at the edge of the area of contact make an included angle of less than $\approx 120°$, and specimens whose ratio of E/Y < 100, lead to deformations of an elastic character[32]. For materials with a higher E/Y, or with a sharper indenter, the mode of deformation appears to be that of radial compression and may be described in terms of the expanding cavity model. It appears that the radial flow pattern observed by Samuels and Mulhearn[21] and given popular attention through the expanding cavity model depends upon the ratio E/Y of the specimen material for a given indenter angle. For conical or Vickers diamond pyramid indenters, the indenter angle is fixed; for a spherical indenter, the effective angle, as measured by tangents to the surface at the point of contact with the specimen, depends on the load.

Let us compare the indentation response of two materials, one with a relatively high value of E/Y, such as mild steel (E/Y = 550), and another with a low value, such as a glass-ceramic (E/Y = 90). Figures 9.3.6 and 9.3.7 show experimental and finite-element results for indentations in these materials using a spherical indenter.

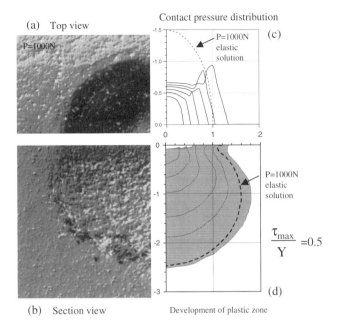

Fig. 9.3.6 Indentation response for glass-ceramic material, E/Y = 90. (a) test results for indenter load of P = 1000 N and indenter of radius 3.18 mm showing residual impression in the surface. (b) Section view with subsurface accumulated damage beneath the indentation site. (c) Finite-element results for contact pressure distribution. (d) Finite-element results showing development of the plastic zone in terms of contours of maximum shear stress at $\tau_{max}/Y = 0.5$. In (c) and (d), results are shown for indentation strains of a/R = 0.035, 0.05, 0.07, 0.09, 0.10, 0.13. Distances are expressed in terms of the contact radius a = 0.315 mm for the elastic case of P = 1000 N.

The predictions of various hardness theories are most markedly characterized by the proposed shape of the plastically deformed region. The expanding cavity model requires a hemispherical plastic zone coincident with the center of contact at the specimen surface. Indeed, such a shape, for metal specimens with spherical and conical or wedge type indenters, has been widely reported in the literature and is demonstrated here in Fig. 9.3.6. However, the hemispherical shape required by the expanding cavity model is not demonstrated for the material with a low value of E/Y as shown in Fig. 9.3.7. In both materials, there is a deviation from linearity in the indentation stress-strain relationship, as shown in Fig. 9.3.8, indicating the presence of plastic deformation within the specimen material.

170   Chapter 9. Hardness

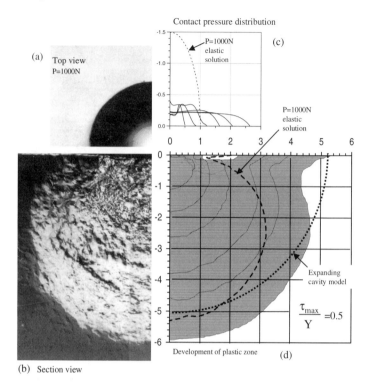

Fig. 9.3.7 Indentation response for mild steel material, E/Y = 550. (a) test results for an indenter load of P = 1000 N and indenter of radius 3.18 mm showing residual impression in the surface. (b) Section view with subsurface accumulated damage beneath the indentation site. (c) Finite-element results for contact pressure distribution. (d) Finite-element results showing development of the plastic zone in terms of contours of maximum shear stress at $\tau_{max}/Y = 0.5$. In (c) and (d), results are shown for indentation strains of a/R = 0.04, 0.06, 0.08, 0.11, 0.14, 0.18. Distances are expressed in terms of the contact radius a = 0.218 mm for the elastic case of P = 1000 N.

Detailed theoretical analysis of events within the specimen material is difficult because of the variable geometry of the evolving plastic zone with increasing indenter load. As load is applied to the indenter, the principal stresses σ_1 and σ_3 within the specimen material increase until eventually the flow criterion is met and thus $|\sigma_1 - \sigma_3| = Y$. An element of such material is shown at (a) in Fig. 9.3.9. Due to the constraint offered by the surrounding elastic continuum, an additional stress σ_R arises, which serves to maintain the flow criterion as the load is increased. Plastic flow occurs until the magnitude of σ_R is such that, with respect to the total state of stress, the net vertical force is sufficient to balance the applied load. Beyond the elastic-plastic boundary, the stresses σ_R diminish until the stress field is substantially the same as the Hertzian elastic case, in accordance with Saint-Venant's principle. Upon removal of load, the elastically strained material attempts to resume its original configuration but is largely prevented from doing so by the plastically deformed material. Except for a slight relaxation due to any elastic recovery that does take place, the stresses σ_R remain within the material and are therefore "residual" stresses (see Section 9.5).

The indentation stress-strain curves in Fig. 9.3.8 show that there is a decrease in the mean contact pressure, compared to the fully elastic case, as plastic deformation occurs beneath the indenter. For the case of a spherical indenter, a decrease in mean contact pressure, at a particular value of indenter load, corresponds to an increase in the size of the contact area and penetration depth. The observed increase in penetration depth indicates an increased energy consumption compared to the fully elastic case since the indenter load does additional work.

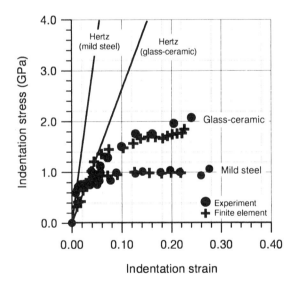

Fig. 9.3.8 Indentation stress-strain curves for materials with a low value of E/Y (glass-ceramic) and high value of E/Y (mild steel). Indentation stress is the mean contact pressure found by dividing the indenter load by the area of contact. Indentation strain is the ratio of the radius of the circle of contact divided by the radius of the indenter. The Hertz elastic solutions for both material types are shown as full lines. Deviation from linearity in the experimental and finite-element data indicates plastic deformation.

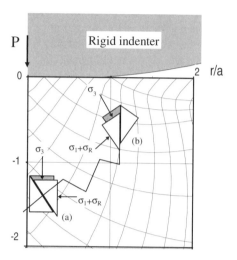

Fig. 9.3.9 Schematic of plastic deformation beneath spherical indenter. Contours of maximum elastic shear stress are drawn in the background. Element of material at (a) has the direction of maximum shear oriented at approximately 45° to the axis of symmetry. Direction of maximum shear follows approximately that of the Hertzian elastic stress field for low value of E/Y. Element of material at (b) undergoes plastic deformation such that the direction of residual field supports the indenter load. Shaded areas indicate plastic strains which are ultimately taken up by elastic strains outside the plastic zone (after reference 31).

Neglecting any frictional or other dissipative mechanisms, it is not immediately evident why there should be more energy transferred from the loading system into strain energy within the specimen material after plastic flow has occurred. It is quite conceivable that, *due to the elastic constraint*, plastic flow occurs and the residual stress field is established without any increase in penetration depth as was thought by Shaw and DeSalvo[24]. However, experimental evidence, in the form of a deviation from linearity on the indentation stress-strain curve, suggests otherwise. The shaded area in Fig. 9.3.9 at (a) indicates the volume of material that is displaced by *additional downward movement* of the indenter as sliding takes place. This displaced volume is accounted for by additional elastic strains in the specimen material within and outside the plastic zone. In Fig. 9.3.9, note that the direction of maximum shear stress for the material at position (a) is approximately 45° to the axis of symmetry and that σ_R acts in a direction normal to the application of load. Thus, due to the orientation of the sliding, the additional elastic strains appear not underneath but off to the side of the plastic zone, where they are less effective in supporting the indenter load. For the material at position (b) in Fig. 9.3.9, similar events occur, but this time the direction of maximum shear is oriented approximately parallel to the direction of applied load. Thus, at this position, the local compliance is increased due to plastic deformation, but a significant component of the residual stress σ_R tends to act in a direction to support the indenter load. These observations account for the shift in the maximum of the contact pressure distribution from the center to the points near the edge of the circle of contact, as shown in Fig. 9.3.7, as plastic deformation proceeds.

What then determines the shape of the plastic zone? For shear driven plasticity, the edge of the plastic zone coincides with the shear stress contour whose magnitude just satisfies the chosen flow criterion. Here it is shown that the location of the edge of the fully developed plastic zone depends on the ratio E/Y. The change in character from a contained to an uncontained plastic zone occurs due to the shift in the balance of elastic strain from material directly beneath the indenter outward toward the edge of the circle of contact. As the plastic zone evolves, material away from the axis of symmetry is being asked to take an increasing level of shear. For materials with a low value of E/Y, a large proportion of this can be accommodated by elastic strain. However, for materials with a high value of E/Y, plastic flow is comparatively more energetically favorable and thus occurs at a lower value of indenter load. The plastic zone thus takes on an elongated shape well before reaching the specimen surface, and the cumulative effect is for the zone to grow ever outward with increasing indenter load. The proximity of the specimen surface also plays a role as the material attempts to accommodate the residual field, and leads to the slight "return" in the shape of the quasi-semicircular plastic zone as shown in Figs. 9.3.6 and 9.3.7. It is thus concluded that the semi-circular plastic zone shape associated with the expanding cavity model and observed in specimens with a high value of E/Y at high values of indentation strain arises due to the nature of the shift in elastic strain energy from material beneath to that adjacent to the evolving plastic zone. The rate of growth of the plastic zone, with respect to increasing indenter load, affects its subsequent shape, the effect being magnified by materials with a high value of E/Y. The distribution of stress around the periphery of the plastic zone becomes more uniform as the gradients associated with the elastic stress field are redistributed as a result of plastic deformation. For both high and low ratios of E/Y, the volume displaced by the indenter is accommodated eventually by elastic strains in the specimen material. As the ratio E/Y increases, the distribution of elastic strain outside the plastic zone assumes a semicircular shape consistent with that required by the expanding cavity model.

References

1. H. Hertz, "On the contact of elastic solids," J. Reine Angew. Math. 92, 1881, pp. 156–171. Translated and reprinted in English in *Hertz's Miscellaneous Papers*, Macmillan & Co., London, 1896, Ch. 5.
2. H. Hertz, "On hardness," Verh. Ver. Beförderung Gewerbe Fleisses 61, 1882, p. 410. Translated and reprinted in English in *Hertz's Miscellaneous Papers*, Macmillan & Co, London, 1896, Ch. 6.
3. A. Wahlberg, "Brinell's method of determining hardness," J. Iron Steel Inst. London, 59, 1901, pp. 243–298.
4. R.L. Smith and G.E. Sandland, "An accurate method of determining the hardness of metals with particular reference to those of high degree of hardness," Proc. Inst. Mech. Eng. 1, 1922, pp. 623–641.
5. F. Knoop, C.G. Peters, and W.B. Emerson, "A sensitive pyramidal-diamond tool for indentation measurements," Research Paper RP1220, National Bureau of Standards, U.S. Dept. Commerce, 1939, pp. 211–240.
6. F. Auerbach, "Absolute hardness," Ann. Phys. Chem. (Leipzig) 43, 1891, pp.61–100. Translated by C. Barus, Annual Report of the Board of Regents of the Smithsonian Institution, July 1, 1890 – June 30 1891, reproduced in "Miscellaneous documents of the House of Representatives for the First Session of the Fifty-Second Congress," Government Printing Office, Washington, D.C., 43, 1891–1892, pp.207–
7. E.Meyer, "Untersuchungen uber Harteprufung und Harte," Phys. Z. 9, 1908, pp. 66–74.
8. S.L. Hoyt, "The ball indentation hardness test," Trans. Am. Soc. Steel Treat. 6, 1924, pp. 396–420.
9. A. Foppl, "Mitteilungen aus dem Mechan," Technische Lab. der Technische Hochschule, Munchen, 1900.
10. C.A. Coulomb, Mem. Acad. Sci. Savants Etrangers, Paris 7, 1776, pp. 343–382.
11. M.S. Paterson, *Experimental Rock Deformation - the Brittle Field*, Springer Verlag, Heidelberg, 1978.
12. C.G. Sammis and M.F. Ashby, "The failure of brittle porous solids under compressive stress states," Acta Metall. 34 3, 1986, pp. 511–526.
13. H. Horii and S. Nemat-Nasser, "Brittle failure in compression: splitting, faulting and brittle-ductile transition," Philos. Trans. R. Soc. London 319 1549, 1986, pp. 337–374.
14. D. Tabor, *The Hardness of Metals*, Clarendon Press, Oxford, 1951.
15. M.C. Shaw, "The fundamental basis of the hardness test," in *The Science of Hardness Testing and its Research Applications*, J.H. Westbrook and H. Conrad, Eds. American Society for Metals, Cleveland, OH, 1973, pp. 1–15.
16. M.V. Swain and J.T. Hagan, "Indentation plasticity and the ensuing fracture of glass," J. Phys. D: Appl. Phys. 9, 1976, pp. 2201–2214.
17. M.T. Huber, Ann. Phys. Chem. 43 61, 1904.
18. R. Hill, E.H. Lee and S.J. Tupper, "Theory of wedge-indentation of ductile metals," Proc. R. Soc. London, Ser. A188, 1947, pp. 273–289.
19. R. Hill, *The Mathematical Theory of Plasticity*, Clarendon Press, Oxford, 1950.

20. D.M. Marsh, "Plastic flow in glass," Proc. R. Soc. London, Ser. A279, 1964, pp. 420–435.
21. L.E. Samuels and T.O. Mulhearn, "An experimental investigation of the deformed zone associated with indentation hardness impressions," J. Mech. Phys. Solids, 5, 1957, pp. 125–134.
22. T.O. Mulhearn, "The deformation of metals by Vickers-type pyramidal indenters," J. Mech. Phys. Solids, 7, 1959, pp. 85–96.
23. K.L. Johnson, "The correlation of indentation experiments," J. Mech. Phys. Sol. 18, 1970, pp. 115–126.
24. M.C. Shaw and D.J. DeSalvo, "A new approach to plasticity and its application to blunt two dimension indenters," J. Eng. Ind. Trans. ASME, 92, 1970, pp. 469–479.
25. M.C. Shaw and D.J. DeSalvo, "On the plastic flow beneath a blunt axisymmetric indenter," J. Eng. Ind., Trans. ASME 92, 1970, pp. 480–494.
26. C. Hardy, C.N. Baronet, and G.V. Tordion, "The Elastic-plastic indentation of a half-space by a rigid sphere," Int. J. Numer. Methods Eng. 3, 1971, pp. 451–462.
27. C.M. Perrott, "Elastic-plastic indentation: Hardness and fracture," Wear 45, 1977, pp. 293–309.
28. S.S. Chiang, D.B. Marshall, and A.G. Evans, "The response of solids to elastic/plastic indentation. 1. Stresses and residual stresses," J. Appl. Phys. 53 1, 1982, pp. 298–311.
29. S.S. Chiang, D.B. Marshall, and A.G. Evans, "The response of solids to elastic/plastic indentation. 2. Fracture initiation," J. Appl. Phys. 53 1, 1982, pp. 312–317.
30. K.L. Johnson, *Contact Mechanics*, Cambridge University Press, Cambridge, U.K., 1985.
31. A.C. Fischer-Cripps, "Elastic-plastic response of materials loaded with a spherical indenter," J. Mater. Sci., 32 3, 1997, pp. 727–736.
32. W. Hirst and M.G.J.W. Howse, "The indentation of materials by wedges," Proc. R. Soc. London, Ser. A311, 1969, pp. 429–444.

Chapter 10
Elastic and Elastic-Plastic Contact

10.1 Introduction

Experiments show that a wealth of information is available concerning the elastic-plastic properties of materials using indentation tests. Having examined elastic and elastic-plastic contact in Chapters 5, 6, 7, and 9, we are now in a position to consider various issues that have a bearing on the interpretation and design of indentation tests. Of particular interest is the connection between different types of indenter and events that occur after the removal of the indenter from the specimen. Depending on the nature of the specimen material and the geometry of the indenter, one may observe brittle cracking of a characteristic pattern or a residual impression which may be either raised up at the edges or sunk down. A very good example is ordinary soda-lime glass. Loading with a spherical indenter usually produces brittle fracture—a conical crack. Loading with a pyramidal indenter yields a plastic residual impression in the specimen surface. These types of phenomena yield information about the mechanical properties of the specimen material.

10.2 Geometrical Similarity

With a diamond pyramid or conical indenter, the ratio of the length of the diagonal or radius of circle of contact to the depth of the indentation[*], d/δ, remains constant for increasing indenter load, as shown in Fig. 10.2.1. Indentations of this type have the property of "geometrical similarity." When there is geometrical similarity, it is not possible to set the scale of an indentation without some external reference.

Unlike a conical indenter, where the ratio d/δ is a constant, the radius of the circle of contact for a spherical indenter increases faster than the depth of the indentation as the load increases. The ratio a/δ increases with increasing load. In this respect, indentations with a spherical indenter are not geometrically similar. Increasing the load on a spherical indenter is equivalent to decreasing the tip

---

[*] In this section only, δ is the indentation depth below the edge of contact, not below the original free surface as in Chapter 5.

semiangle of a conical indenter. However, geometrically similar indentations may be obtained with spherical indenters of different radii. If the ratio a/R is maintained constant, then so is the mean contact pressure, and the indentations are geometrically similar.

The principle of geometrical similarity is widely used in hardness measurements. For example, due to geometrical similarity, hardness measurements made using a diamond pyramid indenter are expected to yield a value for hardness that is independent of the load. For spherical indenters, the same value of mean contact pressure may be obtained with different sized indenters and different loads as long as the ratio of the radius of the circle of contact to the indenter radius, a/R, is the same in each case. The practical importance of such a relationship will be explored further in Chapter 11.

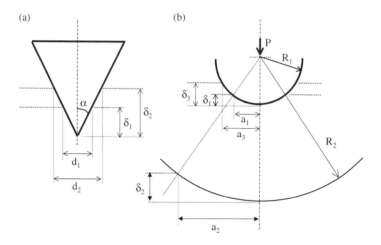

Fig. 10.2.1 Geometrical similarity for (a) diamond pyramid or conical indenter; (b) spherical indenter.

10.3 Indenter Types

10.3.1 Spherical, conical, and pyramidal indenters

Indentation hardness tests are generally made with either spherical, pyramidal, or conical indenters. Consider a Vickers indenter with opposing faces at a semiangle of $\alpha = 68°$ and therefore making an angle $\beta = 22°$ with the specimen surface. For a particular contact radius a, the radius R of a spherical indenter whose edges are at a tangent to the point of contact with the specimen is given by $\sin \beta$ = a/R, which for $\beta = 22°$ gives a/R = 0.375. It is interesting to note that this is precisely the indentation strain[†] at which Brinell hardness tests, using a spherical indenter, are generally performed, and the angle $\alpha = 68°$ for the Vickers indenter was chosen for this reason. The Berkovich indenter[1] is generally used in small-scale indentation studies and has the advantage that the edges of the pyramid are more easily constructed to meet at a single point, rather than the inevitable line that occurs in the four-sided Vickers pyramid. The apex angle of the Berkovich indenter is 65.3°, which gives the same area-to-depth ratio as the Vickers indenter.

Conical indenters have the advantage of possessing axial symmetry and, with reference to Fig. 10.3.1, equivalent projected areas of contact between conical and pyramidal indenters are obtained when:

$$A = 4h^2 \tan^2 \alpha \qquad (10.3.1a)$$

where h is depth of penetration measured from the edge of the circle or area of contact. For $\alpha = 68°$, the projected area of contact is $A = 24.5h^2$ and thus the equivalent angle for a conical indenter is 70.3°.

Now, consider the comparative shapes of the indentation profiles for a sphere and a cone shown in Fig. 10.3.2. If we were to say that the mean contact pressure represents the common ground between different types of indenters, then the shaded volume of material shown underneath the spherical indenter in this figure requires explanation. The indentation depth measured with respect to the edge of the circle of contact for the sphere, h_s, is (from Eq. 6.2.1d):

$$h_s = \frac{a^2}{2R} \qquad (10.3.1b)$$

For small angles of β (large α), then $\sin\beta = a/R = \tan\beta$. But $\tan\beta$ is related directly to the depth h_v beneath the contact circle for the cone. Hence, equating the cone and the sphere, we obtain:

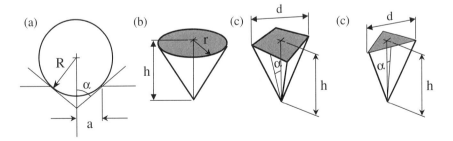

Fig. 10.3.1 Indentation parameters for (a) spherical, (b) conical (c) pyramidal, and (d) Berkovich indenters.

[†] Recall that the term "indentation strain" refers to the ratio a/R.

Chapter 10. Elastic and Elastic-Plastic Contact

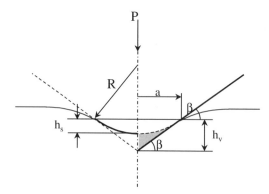

Fig. 10.3.2 Comparative geometries for indentation with a sphere and cone.

$$\tan \beta = \frac{a}{R}$$

$$\frac{a}{R} = \frac{h_v}{a}$$

$$\frac{2h_s}{a} = \frac{h_v}{a} = \frac{a}{R} \qquad (10.3.1c)$$

$$h_s = \frac{1}{2} h_v$$

Now, for a given value of load P leading to equivalent contact areas of radius a (and hence identical mean contact pressures), it is evident that more energy is required to obtain this contact pressure for the case of the cone since more material has to be displaced by the indenter. Equation 6.2.3c in Chapter 6 shows that, for the case of the cone, the indentation depth $h = u_{z|r=0}$ is proportional to $P^{1/2}$ and for the sphere Eq. 6.2.1i shows that the indentation depth is proportional to $P^{2/3}$. Thus, although the mean contact pressures may be the same, the work done in achieving this contact pressure is higher for the case of the cone than for the sphere since, on a plot of P vs h, the rate of increase of P with h is initially higher for the cone than for the sphere.

During an indentation stress-strain test involving a spherical indenter, one may be tempted to conclude that the limiting value of the mean contact pressure is the hardness value H. That is, within Region 3, there is no difference in the mean contact pressure obtained with a Vickers diamond pyramid indenter and that obtained with a spherical indenter. This infers that the constraint factor is the same for each type of indenter. Although experimental evidence suggests that this is approximately the case, there is no particular physical reason for this behavior. It is interesting to note that the comparison made by Samuels and Mulhearn[2] is usually presented without regard to the different distance scales on the vertical axes in their diagrams. However, if the size of the plastic zone is large in comparison to the size of the radius of circle of contact, then one may consider the mean contact pressure to be independent of the shape of the indenter.

10.3.2 Sharp and blunt indenters

Indenters can generally be classified into two categories—sharp or blunt. The criteria upon which a particular indenter is classified, however, are the subject of opinion. For example, some authors[3] classify sharp indenters as those resulting in permanent deformation in the specimen upon the removal of load. A Vickers diamond pyramid is such an example in this scheme. However, others[2] prefer to classify a conical or pyramidal indenter with a cone semiangle $\alpha > 70°$ as being blunt. Thus, a Vickers diamond pyramid with $\alpha = 68°$ would in this case be considered blunt. A spherical indenter may be classified as sharp or blunt depending on the applied load according to the angle of the tangent at the point of contact. The latter classification is based upon the response of the specimen material in which it is observed that plastic flow according to the slip-line theory occurs for sharp indenters and the specimen behaves as a rigid-plastic solid. For blunt indenters, the response of the specimen material follows that predicted by the expanding cavity model or the elastic constraint model, depending on the type of specimen material and magnitude of the load. Generally speaking, cylindrical flat punch and spherical indenters are termed blunt, and cones and pyramids are sharp.

10.4 Elastic-Plastic Contact

10.4.1 Elastic recovery

Consider an element of material, surrounded by an elastic continuum, and loaded by a compressive stress σ_1 as shown in Fig. 10.4.1 (a). Since in this figure $\sigma_3 = 0$, the maximum shear stress $\tau_{max} = \sigma_1/2$. Assuming the Tresca criterion for plastic flow is appropriate ($\sigma_1 - \sigma_3 = Y$), slippage occurs when $\sigma_1 = Y$ and a section of the element slides downward as shown in Fig. 10.4.1 (b). However, this sliding section is constrained by a surrounding elastic continuum. The side of the section acts as if to "indent" the elastic surroundings, and a reaction stress σ_R is created as a result The magnitude of σ_R increases as σ_1 increases so as to keep the flow criterion satisfied within the element. The yield stress Y is a measure of the cohesive strength of the sliding interface. At full load, within the element, both elastic and plastic strains exists. In the elastic continuum, only elastic strains exist. Due to the constraint offered by the elastic continuum, the plastic strains are restricted to an order of magnitude comparable to the elastic strains. The total state of stress is given by the superposition of σ_1 and σ_R. Since plastic deformation has occurred at this point, this total distribution of stress may be termed the "elastic-plastic" stress field σ_{ep}.

When the load σ_1 is removed, the element and the elastic continuum attempt to regain their original configuration but are prevented from doing so by the permanently deformed element. The stresses σ_R thus remain acting on the deformed element and are called "residual" stresses. Since the element is subjected to the stress σ_R only, the flow criterion then becomes $\sigma_R = Y$, which, if satisfied, may cause "reverse" plasticity. Relaxation of the initial elastic strains within and outside the element upon removal of the load σ_1 results in a partial resumption in shape of the material which is termed "elastic recovery," as shown in (d) in Fig. 10.4.1. If the stress σ_1 is now reapplied, then the element resumes the shape which it took at full load (i.e., back to condition (c) in Fig. 10.4.1) and the elastic-plastic stress field is re-established. The reapplication of load involves only elastic displacements. However, this reloading, even though it may be a completely elastic process, is not energetically equivalent to the loading of an initially stress-free elastic material, since the system is already in a state of residual stress. That is, more work is required in compressing a spring if it is pre-compressed compared with the same deformation from its free length. The "pre-compression" in this case is the residual stress σ_R.

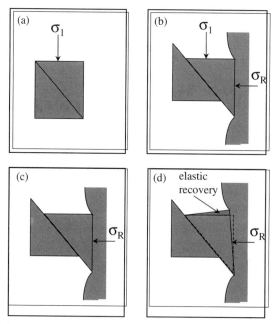

Fig. 10.4.1 Elastic constraint and residual stresses. (a) compressive loading with no elastic constraint; (b) compressive loading with elastic constraint; (c) elastic and plastic deformation at full load; (d) removal of load and residual stresses.

Consider now the profile of the fully loaded and unloaded indentation impressions shown in Fig. 10.4.2 (a). After unloading, elastically strained material outside the plastic zone will attempt to resume its original shape but is mostly prevented by permanently deformed plastic material beneath the indenter. The specimen material is therefore left in a state of residual stress. Any resulting differences in the shape of the indentation profile between that at full load and unload will be due to *elastic* recovery of the specimen material. Experiments show that the radius of circle of contact between fully loaded and fully unloaded conditions remains virtually unchanged due to geometry of the indentation loading. Reloading the indenter involves purely elastic deformation of the elastically recovered profile. The distribution of contact pressure required to bring the indentation profile back to that at full load is that which exists beneath the indenter at full load (i.e., the elastic-plastic pressure distribution)[‡]. As with any elastic contact, the elastic-plastic pressure distribution can thus be determined from the superposition of line or point loads where the total elastic displacement u_z is the sum of the component displacements at a particular radius r (see Section 5.3.3):

$$u_z = \int p(r) dr \qquad (10.4.1a)$$

The total displacement u_z at any particular value of r may be determined experimentally. This is particularly straightforward if the indenter is considered to be rigid since the displacement of the surface at full load simply matches the profile of the indenter. The shape, or profile, of the residual impression may also be measured experimentally. The differences between the loaded and unloaded configurations are thus the elastic displacements, h_e, which occur during reloading. Working backward from Eq. 10.4.1a would provide the pressure distribution p(r), which would result in these displacements. Hirst and Howse[4] used this pressure distribution as being representative of that associated with the elastic-plastic loading (i.e., the elastic-plastic pressure distribution at full load). However, as noted previously, the residual stresses σ_R impose a "pre-stress" condition on these elastic displacements which is not accounted for in this procedure. The error was probably not significant for Hirst and Howse since their perspex specimens exhibited elastic recoveries on the order of ≈75%.

The occurrence of reverse plasticity during unloading depends on the magnitude of σ_R and the yield stress Y. The term "elastic recovery" should be used with care since it refers to a purely elastic event involving no reverse plasticity. Reloading of an indenter should be thought of as generally involving both elastic deformations as well as a partial repeat of the original elastic-plastic contact to account for the possibility of reverse slip. Figure 10.4.3 shows the result of repeated indentations, compared to a single application of load, for a glass-ceramic. Note that the degree and severity of damage are markedly increased

---

[‡] This result applies to both elastic and elastic-plastic deformation of the specimen material.

after many repeated applications of load, indicating that the reloading, in this particular material, is not at all an elastic event and that reverse slip must occur upon unloading.

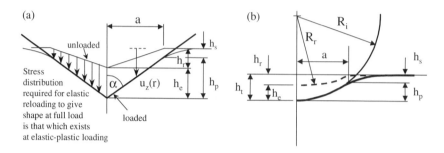

Fig. 10.4.2 Schematic of elastic-plastic indentation with (a) conical indenter and (b) spherical indenter. For a perfectly rigid indenter, h_t is the total indentation depth, h_r is the depth of the residual impression, h_e is the indentation depth associated with the elastic unloading/reloading, and h_p is the depth of the circle of contact at full load.

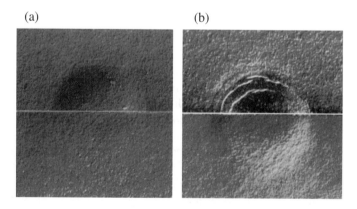

Fig. 10.4.3 Indentation contact for a glass ceramic. (a) Single application of load; (b) repeated application of load. The top half of each figure shows the impression in the specimen surface, and the bottom half shows a section view through the thickness of the specimen beneath the circle of contact.

10.4.2 Compliance

When plasticity occurs during an indentation, an increase in indentation depth over that expected for a purely elastic deformation is generally observed. Consider the case of a cylindrical flat punch indenter. For this type of indenter, the indentation depth, h, beneath the specimen free surface is given by (see Section 6.2.2):

$$h = \frac{(1-\nu^2)}{2E}\frac{P}{a} \qquad (10.4.2a)$$

A schematic of the indenter load P vs indentation depth h is shown in Fig. 10.4.4. For the fully elastic case, the response of the material upon loading and unloading is the straight line segment AC. When full load is applied, the elastic, recoverable strain energy is the area bounded by ACF. For the elastic-plastic case, loading proceeds along the curved path ABD. For elastic unloading, loading proceeds from D to G. If there were no residual stresses, or complete elastic recovery, then unloading would proceed from D to A. The area bounded by ABD is thus an indication of energy dissipated during plastic deformation. The area bounded by the triangle ADG is then an indication of the energy stored within the residual field.

Shaw and DeSalvo[5] based their elastic constraint model (see Chapter 9) on the assumption that the establishment of the residual field requires no additional energy beyond that required for the fully elastic case (i.e., path ACH in Fig. 10.4.4). However, the energy required to establish the residual field is that indicated by the area ADG and that dissipated by plastic events by the area ABD. Although Fig. 10.4.4 applies to a cylindrical indenter, the same general principles apply to the case of a spherical indenter.

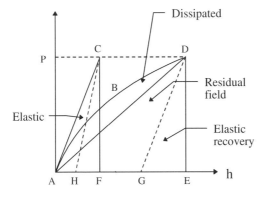

Fig. 10.4.4 Schematic of load versus indentation depth for cylindrical flat punch indenter.

10.4.3 Analysis of compliance curves

10.4.3.1 Submicron indentation testing

The study of the indentation response of thin films requires indentations on a very small scale, usually in the submicron regime. A three-sided Berkovich indenter is often employed for this type of work in preference to a four-sided Vickers indenter. This is because the three sides of the Berkovich indenter meet at a single point rather than the line that is usually observed at the tip of a Vickers indenter. Equipment for this purpose usually consists of an instrumented loading device that records the indenter load in mN and displacement in microns or even nanometers. Estimations of both elastic modulus and hardness of the specimen material are possible from load versus penetration measurements. Rather than a direct measurement of the size of residual impressions, which may require electron microscopy, contact areas are actually calculated from depth measurements together with a knowledge of the shape of the indenter. This is in contrast to the procedure used for large-scale indentation experiments, where the lateral size, rather than the depth, of the residual impression in the specimen surface is used as the basis for calculating hardness. The following sections describe the method by which elastic modulus and hardness are obtained from indenter load and depth measurements. These methods are undergoing continuing development as workers in the field apply corrections to account for material behavior, such as piling up or sinking in, and imperfections in indenter geometry.

10.4.3.2 Cylindrical indenter

Consider the case of a cylindrical flat punch indenter that has an elastic-plastic load displacement response, as shown by the path ADG in Fig. 10.4.4 and is reproduced in Fig. 10.4.5.

Figure 10.4.5, (a) shows the displacements for the elastic-plastic contact at full load P_{max} and the displacements at full unload. The unloading response is assumed to be fully elastic. Elastic displacements can be calculated using Eq. 10.4.2a. With h equal to the displacement u_z at r = 0, and by taking the derivative dP/dh, we can arrive at an expression for the slope of the unloading curve:

$$P = \frac{2aE}{1-v^2} h$$
$$\frac{dP}{dh} = \frac{2aE}{1-v^2}$$

(10.4.3.2a)

In Eq. 10.4.3.2a, a is the contact radius, which, for the case of a cylindrical punch, is equal to the radius of the indenter. Expressing this in terms of the contact area:

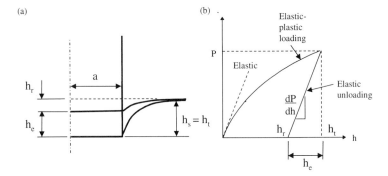

Fig. 10.4.5 (a) Schematic of indenter and specimen surface geometry at full load and full unload for cylindrical punch indenter. (b) Load versus displacement for elastic-plastic loading followed by elastic unloading. h_r is the depth of the residual impression, h_t is the depth from the original specimen surface at maximum load P_{max}, h_e is the elastic displacement during unloading, and h_s is the distance from the edge of the contact to the specimen surface, here equal to h_t for cylindrical indenter.

$$\frac{dP}{dh} = \frac{2E\sqrt{A}}{\sqrt{\pi}(1-v^2)} \qquad (10.4.3.2b)$$

Pharr, Oliver, and Brotzen[6] show that Eq. 10.4.3.2b applies to all axis-symmetric indenters. Equation 10.4.3.2b shows that the slope of the unloading curve is proportional to the elastic modulus and may be calculated from the known radius of the punch. As shown in Fig. 10.4.5, h_e is the displacement for the elastic unloading. Thus, the slope of the unloading curve is also given by:

$$\frac{dP}{dh} = \frac{P_{max}}{h_e} \qquad (10.4.3.2c)$$

Now, for a cylindrical indenter, there is no need for an estimation of the size of the contact area from depth measurements since it is equal to the radius of the indenter. However, the situation becomes quite complicated when this is not the case, such as for the Berkovich indenter.

10.4.3.3 Pointed indenters

As can be seen from Fig. 10.4.5, the slope of the elastic unloading curve for a cylindrical punch indenter is linear. Doerner and Nix[7] observed that for tests with a Berkovich indenter, the *initial* unloading curve is linear for a wide range of test materials. These workers then applied cylindrical punch equations to the initial part of an unloading curve to determine the size of the contact from depth

188   Chapter 10. Elastic and Elastic-Plastic Contact

measurements. Their analysis considered the case of a conical indenter and assumed that the nonaxis-symmetric nature of the pyramid had little effect on the final result[6].

Consider the elastic-plastic loading and elastic unloading of a specimen with a conical indenter. The shape of the surface for the sequence is shown in Fig. 10.4.6 (a).

As the indenter is unloaded from full load, the contact radius remains fairly constant (due to a fortuitous combination of the geometry of the deformation and the shape of the indenter) until the surface of the specimen no longer conforms to the shape of the indenter. Thus, for the initial part of the unloading, if the contact radius is assumed to be constant, the unloading curve is linear. Unloading from the fully loaded impression for a cone is therefore similar to that seen for the elastic unloading of a cylindrical punch. Doerner and Nix[7] thus use Eq. 10.4.3.2b, with A being the contact area of the punch, to obtain the depth of the edge of the circle of contact h_s, and hence h_p. The radius of the circle of contact at full load is obtained from the slope of the initial unloading.

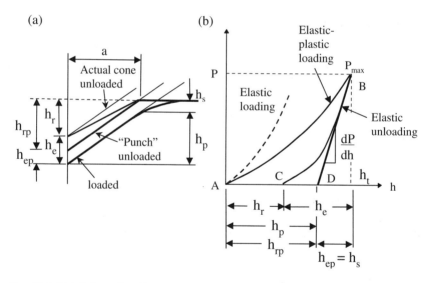

Fig. 10.4.6 (a) Schematic of indenter and specimen surface geometry at full load and full unload for conical indenter. (b) Load versus displacement for elastic-plastic loading followed by elastic unloading. h_r is the depth of the residual impression, h_{rp} is the depth of the residual impression for an equivalent punch, h_t is the depth from the original specimen surface at maximum load P_{max}, h_e is the elastic displacement during unloading of the actual cone, h_{ep} is the elastic displacement for an equivalent punch, and h_s is the distance from the edge of the contact to the specimen surface at full load.

With reference to Fig. 10.4.6 (b), for the initial unloading, the cone acts like a punch in terms of its "constant" area of contact, since the initial unloading is appears to be linear. If the area of contact remained constant during the entire unloading, then the unloading curve would be linear: from P_{max} to $P = 0$. We could then ask ourselves, "What would be the unloading curve associated with a punch that has the slope dP/dh at P_{max}?" The answer would be that which is extrapolated to zero load; that is, the path BD in Fig. 10.4.6 (b). If unloading were to take place along this line, then the elastic displacement of this imaginary punch would be the distance h_{ep} (the subscript "p" denoting "punch"). The cone actually travels the path BC during unloading, which is a distance h_e, leaving a residual impression of depth h_r. Now, with reference to Fig. 10.4.6 (a), it is easy to see that for a cone that acts like a punch (by having a constant radius of circle of contact during unloading) and unloads through a distance h_{ep}, the distance h_{ep} is equal to the distance h_s *that exists at full load*, which is where we require an estimate of h_s for the actual cone. As a consequence, the distance h_{rp}, which is the intercept of the unloading curve for the punch, is equal to h_p. Thus:

$$h_t - h_{rp} = h_s$$
$$h_{rp} = h_t - h_s \quad (10.4.3.3a)$$
$$= h_p$$

Equation 10.4.3.3a indicates that the depth h_{rp}, and hence h_p, can be obtained from the intercept of the linear unloading curve with the displacement axis. Once h_p is known, then the area of the contact can be calculated and the hardness and elastic modulus determined from the geometry of the indenter. For example, for a Vickers or Berkovich indenter, the relationship between the projected area A_p of the indentation and the distance h_p is:

$$A_p = 24.5 h_p^2 \quad (10.4.3.3b)$$

Thus:

$$\frac{dP}{dh} = \frac{2h_p E}{1-v^2} \sqrt{\frac{24.5}{\pi}} \quad (10.4.3.3c)$$

The displacement h_p is found from the intercept of the linear unloading curve with the displacement axis. The procedure assumes that the initial unloading is linear. The actual cone and the imaginary punch meet at P_{max}. We assume that the initial unloading can be extrapolated back to zero load, which gives us a measure of h_{ep}, and then we say that due to the geometrical similarity of this punch-like cone $h_{ep} = h_s$ and thus we find h_p for the actual cone. Values of H and E can be calculated from the maximum load P_{max} divided by the projected area (Eq. 10.4.3.3b) A_p and from Eq 10.4.3.3c, respectively.

This imaginative solution to a complex problem was investigated more thoroughly by Oliver and Pharr[8]. These workers observed that for many materials the slope of the initial unloading curve was not linear but slightly curved (i.e., more like the elastic unloading of a cone rather than a punch). This observation led these investigators to formulate an improved method of analysis which is described below.

For completely elastic contact, the relationship between the load and the depth of penetration for a cone is given by Eq. 6.2.3a, and is repeated here for convenience:

$$P = \frac{\pi a E}{2(1-v^2)} a \cot \alpha \qquad (10.4.3.3d)$$

where the quantity a cot α is the depth of penetration measured at the circle of contact, ie. h_p. Thus:

$$P = \frac{\pi a E h_p}{2(1-v^2)} \qquad (10.4.3.3e)$$

Now, from Eq. 6.2.3c, the load P can be expressed in terms of an elastic displacement h measured from the specimen free surface as:

$$P = \frac{2E \tan \alpha}{\pi(1-v^2)} h_{r=0}^2 \qquad (10.4.3.3f)$$

and the slope of the elastic unloading given by:

$$\frac{dP}{dh} = 2 \frac{2E \tan \alpha}{\pi(1-v^2)} h_{r=0} \qquad (10.4.3.3g)$$

Substituting back into Eq. 10.4.3.3f, we have:

$$P = \frac{1}{2} \frac{dP}{dh} h_{r=0} \qquad (10.4.3.3h)$$

Equation 10.4.3.3g shows that the slope of the elastic unloading for a cone is not linear but depends on h. As shown in Fig. 10.4.7, as the indenter is unloaded, then the tip of the indenter moves through a distance h_e and the edge of the circle of contact with the specimen surface moves through a distance h_s. Also, at full load, a cot $\alpha = h_p$. The actual unloading occurs along the line BC in Fig. 10.4.7 (b), but there is at first no information regarding exactly where along this line we might obtain a value for h_s. However, if we regard the elastic unloading of a cone, rather than a punch, through a total distance h_e, then at $P = P_{max}$ the displacements $h_{r=0} = h_e$ and $h_{r=a} = h_s$ and a cot $\alpha = h_p$. Making use of Eq. 6.2.3b, and inserting r=0 and r=a, respectively, we have:

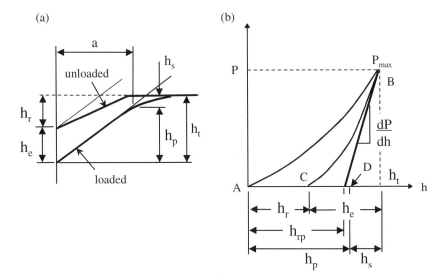

Fig. 10.4.7 (a) Schematic of indenter and specimen surface geometry at full load and full unload for conical indenter. (b) Load versus displacement for elastic-plastic loading followed by elastic unloading. h_r is the depth of the residual impression, h_{rp} is the depth of the residual impression for an equivalent punch, h_t is the depth from the original specimen surface at maximum load P_{max}, h_e is the elastic displacement during unloading of the actual cone, and h_s is the distance from the edge of the contact to the specimen surface at full load. Upon elastic reloading, the tip of the indenter moves through a distance h_e, and the eventual point of contact with the specimen surface moves through a distance h_s.

$$h_e = \frac{\pi}{2} h_p$$
$$h_s = \left(\frac{\pi}{2} - 1\right) h_p \qquad (10.4.3.3i)$$

and thus:

$$h_s = \left(\frac{\pi - 2}{\pi}\right) h_e \qquad (10.4.3.3j)$$

But, from Eq. 10.4.3.3h, at P_{max} and $h_{r=0} = h_e$, we have:

$$h_s = \left[\frac{2(\pi - 2)}{\pi}\right] \frac{P_{max}}{dP/dh} \qquad (10.4.3.3k)$$

Equation 10.4.3.3k should be compared with Eq. 10.4.3.2c. It is indeed fortuitous that these two equations are identical except for the square bracketed term which we might regard as a "correction factor," equal to 0.72, applied to the cylindrical punch equation to account for the unloading of a cone. The significance of this is that we can now use the intercept of a line drawn tangent to the unloading curve at $P = P_{max}$ (which we would have done if we were using the cylindrical punch treatment), adjusted by a correction factor to obtain a value for h_s and hence h_p for the actual unloading of a cone. The correction factor is simply 1 for the case of a real cylindrical punch. It is easy to show that the value of h_p so obtained is a little larger than that obtained from the cylindrical punch approximation. Oliver and Pharr[8] undertook a series of experiments to test this analysis and found that the correction factor to be applied should be 0.75, somewhat higher than that calculated for the cone. This larger figure accounts for plastic deformation within the specimen material which occurs in the vicinity of the tip and causes the pressure distribution over the contact area to resemble that which would occur during an indentation with a cone with a slightly blunted tip.

There is also some slight error associated with the nonaxis-symmetric nature of these indenters with respect to the analysis given above, but it appears to be on the order of about 1–2%[6]. Recent literature in this area is concerned with the applicability of this procedure to materials that display appreciable piling up and sinking in[9,10], the influence of indenter tip radius[11,12], and the effects of work hardening and yield[13] on the elastic modulus.

In the analysis above, we have not taken into consideration the compliance of the indenter (contained within the term E^* in Chapter 6) or the measuring instrument. Further, the shape function for a particular indenter requires experimental calibration to account for geometrical imperfections.

10.4.3.4 Spherical indenter

Field and Swain[14] have developed a compliance analysis for submicron indentations with a spherical indenter. Figure 10.4.8 (b) shows a schematic of load P versus displacement $h = u_{z|r=0}$ for a spherical indenter. Both loading and unloading curves are shown. Assuming a perfectly elastic indenter, the difference between the two curves is an indication of plastic deformation in the specimen during loading.

For an elastic contact, the load-point displacement δ is given by Eq. 6.2.1i, repeated here for convenience (recall that the load-point displacement δ is equivalent to the indentation depth $u_{z|r=0}$ for a rigid indenter of radius R_i and E^* is an equivalent modulus of the system—see Eq. 6.2.1h).

$$\delta^3 = \left(\frac{3}{4E^*}\right)^2 \frac{P^2}{R_i} \qquad (10.4.3.4a)$$

Now, with reference to Fig. 10.4.8 (b), let us assume that upon increasing load, there is an initial elastic deformation followed by a plastic response. The depth of penetration beneath the original specimen free surface is h_t at full load P_t. When the load is removed, assuming no reverse plasticity, the unloading is elastic and at complete load, there is a residual impression of depth h_r. If the load P_t is then reapplied, then the reloading is elastic and takes the form of Eq. 10.4.3a through a distance $h_e = h_t - h_r$ according to:

$$P = \frac{4E^*}{3} R^{\frac{1}{2}} h_e^{3/2} \qquad (10.4.3.4b)$$

However, this elastic reloading involves the deformation of the preformed residual impression as shown in Fig. 10.4.8 (a). Thus, R in Eq. 10.4.3b is the relative radius of curvature of the preformed impression R_r and the indenter R_i given by:

$$\frac{1}{R} = \frac{1}{R_i} - \frac{1}{R_r} \qquad (10.4.3.4c)$$

There are two important matters to consider at this point. The size of the residual impression may be assumed to be identical to that of the radius of the circle of contact at full load. That is, during reloading of the residual impression, the point of contact between the indenter and the specimen moves outward (and downward) until it meets the edge of the residual impression, by which time, the force has reached P_{max}. Since the loading from h_r to h_t is elastic, Eq. 6.2.1d shows that the depth of the circle of contact beneath the specimen free surface is half of the elastic displacement h_e. That is, the distance from the specimen free surface (at full unload) to the depth of the radius of the circle of contact at full load is $h_e/2$. With reference to Fig. 10.4.2:

$$h_t = h_p + \frac{h_e}{2}$$
$$h_p = h_t - \frac{h_e}{2} \qquad (10.4.3.4d)$$
$$= h_t - \frac{h_t - h_r}{2} = \frac{h_t + h_r}{2}$$

From geometry, the radius of the circle of contact at full load (at h_p) is given by:

$$a = [2R_i h_p - h_p^2]^{1/2} \qquad (10.4.3.4e)$$

The area of indentation can be found from:

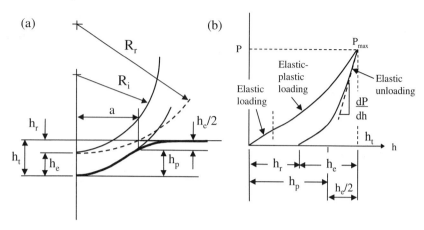

Fig. 10.4.8 (a) Geometry of loading a preformed impression of radius R_r with a rigid indenter radius R_i. (b) Compliance curve (load versus displacement) for an elastic-plastic specimen loaded with a spherical indenter showing both loading and unloading response. Upon loading, there is an initial elastic response followed by elastic-plastic deformation. Upon complete unload, there is a residual impression of depth h_r.

$$A = \pi[2R_i h_p - h_p^2] \quad (10.4.3.4f)$$

The hardness H is thus:

$$H = \frac{P}{\pi(2R_i h_p - h_p^2)} \quad (10.4.3.4g)$$

and the modulus (see Eqs. 6.2.1g and 6.2.1i):

$$E^* = \frac{3}{4} \frac{P}{(2R_i h_p - h_p^2)^{\frac{1}{2}} (h_t - h_r)} \quad (10.4.3.4h)$$

The only unknown in the above equations is the depth of the residual impression h_r. Although this may be measured experimentally by unloading the indenter and recording the depth at which the load drops to zero, this procedure cannot guarantee a purely elastic response since residual stresses may lead to reverse plasticity within the specimen material. An alternative is to partially unload the indenter to a load P_s from the full load P_t. The measuring instrument usually provides values of P and h directly. Thus, P_t, h_t, P_s and h_s are measurable quantities. Following Field and Swain[14], the depth h_r is found from:

$$h_e = h_t - h_r = \left[\left(\frac{3}{4E^*}\right)^{\frac{2}{3}} \frac{1}{R^{\frac{1}{3}}}\right] P_t^{2/3}$$

$$h_s - h_r = \left[\left(\frac{3}{4E^*}\right)^{\frac{2}{3}} \frac{1}{R^{\frac{1}{3}}}\right] P_s^{2/3} \qquad (10.4.3.4i)$$

$$\frac{h_t - h_r}{h_s - h_r} = \left(\frac{P_t}{P_s}\right)^{\frac{2}{3}}$$

$$h_r = \frac{h_s (P_t/P_s)^{2/3} - h_t}{(P_t/P_s)^{2/3} - 1}$$

Values for E^* and H can now be calculated from experimental measurements using Eqs. 10.4.3.4g and 10.4.3.4h. It is important to remember that in the above derivations, we have regarded the spherical indenter to be perfectly rigid although E for the specimen can still be extracted from Eq. 6.2.1h.

We have not required an explicit value for the radius R_r in the above treatment, but this can be calculated from geometry using Eq. 10.4.3.4f. That is:

$$R_r = \frac{a^2 + h_r^2}{2h_r} \qquad (10.4.3.4j)$$

Equation 10.4.3.4j calculates the radius of residual impression R_r that corresponds to the chordal depth h_r beneath the specimen free surface. This radius should also be calculable from Eq. 10.4.3.4b with $P=P_t$ and $h=h_e$. Rearranging Eq. 10.4.3.4b and making use of Eq. 6.2.1g, we obtain an expression for R_r in terms of a, the radius of circle of contact at full load, R_i, P_t and E:

$$R_r = \frac{R_i a^3}{a^3 - R_i P_t \frac{3}{4E^*}} \qquad (10.4.3.4k)$$

Equation 10.4.3.4k calculates the radius of residual impression which needs to be elastically loaded by a rigid indenter R_i at a load P_t to give a radius of circle of contact equal to a. Calculations show that the value for R_r given by Eq. 10.4.3.4k is typically smaller than that given by Eq. 10.4.3.4j. The discrepancy arises because of the limitations of the Hertz equations. It will be recalled from Chapter 6 that it is assumed that the radii of curvature of the contacting bodies are large compared with the radius of the circle of contact. This may not always be the case in actual indentation testing, particularly in the submicron regime, where indentation strains of up to 0.5 are commonly encountered.

10.4.4 The elastic-plastic contact surface

In Chapter 6, it was shown that for an elastic contact between two spheres, the profile of the surface of contact is spherical, with a radius of curvature intermediate between that of the contacting bodies and more closely resembling that body with the greatest elastic modulus. Thus, contact between a flat surface and a nonrigid indenter of radius R is equivalent to that between the flat surface and a perfectly rigid indenter of a larger radius R^*, which may be computed using Eq. 6.1a with k set as for a rigid indenter. Of further interest is the profile of the deformed surfaces when plastic deformation occurs in the specimen and/or the indenter material. Finite-element analysis[15] shows that there is, not surprisingly, a considerable increase in the indentation depth over that for an elastic contact since, due to plastic deformation, the same indenter load now must be supported by strains elsewhere in the specimen material rather than immediately beneath the indenter. Figure 10.4.9 shows the displacements of the specimen surface for an elastic-plastic specimen response.

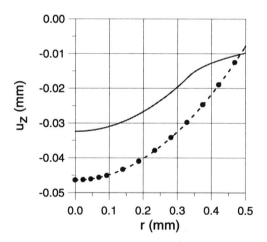

Fig. 10.4.9 Displacements of the specimen surface in the vertical, or z axis, direction in the vicinity of the contact surface at full load (P = 1000 N, R = 3.18 mm) for specimen elastic-plastic response. Solid line indicates Hertzian theoretical result. Data points indicate finite-element results, and dashed line is the spherical contact surface computed using fitted values of R^* from finite-element results (after reference 15).

Table 10.4.1 Comparison between finite-element results for a WC indenter and an elastic-plastic glass-ceramic specimen and theoretical results for purely elastic contact with P = 1000 N and R = 3.18 mm.

	a (mm)	u_z (mm)	δ (mm)	R^* (mm)
Hertz	0.3375	−0.0325	0.0358	3.518
Elastic-plastic F.E. result	0.4766	−0.0464	0.0485	3.257±2.2×10^{-5}
Δ%	+29.2	+50.6	+26.1	−8.0

The finite-element results suggest that the indentation surface does have a spherical profile, as indicated by the very low uncertainty in the fitted value of the effective radius R^*. Table 10.4.1 shows that although there is a considerable discrepancy between the effective radius R^* as calculated using Eq. 6.2.1j (as would be expected since this equation assumes no plastic deformation), the results show that the profile of the surface beneath the indenter is spherical but has a smaller radius of curvature.

10.5 Internal Friction and Plasticity

Plasticity is usually associated with slippage across planes within the material due to shear stresses. Slippage between such planes raises the question of the effect of "internal" friction. Jaeger and Cook[16] show that slip along an internal plane of weakness may occur when the Coulomb[17] criterion is satisfied:

$$|\tau| = S_o + \mu \sigma_N \qquad (10.5a)$$

In Eq. 10.5a, τ is the applied shear stress, S_o is a constant for the material, μ is the coefficient of "internal" sliding friction, and σ_N is the normal stress on the plane under consideration. The parameter S_o represents the inherent shear strength of the material. The normal stress σ_N and shear stress τ on a plane oriented at angle β can be expressed in terms of the maximum shear stress τ_{max} and mean, or average, stress σ_m.

$$\sigma_N = \sigma_m + \tau_{max} \cos 2\beta$$
$$\tau = -\tau_{max} \sin 2\beta \qquad (10.5b)$$

where

$$\tau_{max} = \frac{|\sigma_1 - \sigma_3|}{2}$$

$$\sigma_m = \frac{(\sigma_1 + \sigma_3)}{2}$$

For $\beta = 45°$, $\tau = \tau_{max}$ and $\sigma_N = \sigma_m$. Thus, for the two-dimensional case, the criterion for failure, or sliding, is:

$$(\mu^2 + 1)^{1/2} \tau_{max} - \mu\sigma_m = S_o \tag{10.5c}$$

or

$$\sigma_1\left[(\mu^2 + 1)^{1/2} - \mu\right] - \sigma_3\left[(\mu^2 + 1)^{1/2} + \mu\right] = 2S_o \tag{10.5d}$$

In Eq. 10.5d and those to follow, we specify the *positive* value of σ_1 as being the maximum value of *compressive* stress in accordance with the rock mechanics literature (i.e., σ_1 here is equal to $-\sigma_3$ in conventional terminology). The Tresca failure criterion (Eq. 1.4.1) is thus a special case of the Coulomb criterion (Eq. 10.5d) with $\mu = 0$ and $S_o = Y/2$.

The angle at which sliding occurs is found from:

$$\tan 2\beta = -\frac{1}{\mu} \tag{10.5e}$$

and thus β is between 45° and 90°. For the case of $\mu = 0$, $\beta = 45°$. In real materials, where $\mu \neq 0$, shear fracture does not take place in the direction of maximum shear stress but, due to friction, is always inclined at an acute angle to the direction of maximum compressive stress σ_1.

Equations 10.5a to 10.5e were developed by Coulomb[17] and are given by Jaeger and Cook[16] for shear failure in rock, but the same principles apply to the indentation experiments. Consider a simple uniaxial compression with $\sigma_3 = 0$. We may call the value of σ_1 at which the failure criterion is met the uniaxial compressive strength C_o where, from Eq. 10.5e:

$$C_o = 2S_o\left[(\mu^2 + 1)^{1/2} + \mu\right] \tag{10.5f}$$

Substituting back into Eq. 10.5.3b gives the following flow criterion:

$$\sigma_1 = \left[(\mu^2 + 1)^{1/2} + \mu\right]^2 \sigma_3 + C_o \tag{10.5g}$$

The friction coefficient μ can be determined from the slope of a plot of σ_1 vs σ_3 at failure and the inherent shear strength from the σ_1 axis intercept. According to Eq. 10.5g, when plasticity occurs, the relationship between σ_1 and σ_3 should be linear if the material follows the Coulomb criterion.

References

1. E.S. Berkovich, "Three-faceted diamond pyramid for micro-hardness testing," Ind. Diamond Rev. $\underline{11}$ 127, 1951, pp. 129–133.
2. L.E. Samuels and T.O. Mulhearn, "An experimental investigation of the deformed zone associated with indentation hardness impressions," J. Mech. Phys. Solids, $\underline{5}$, 1957, pp. 125–134.
3. B.R. Lawn, *Fracture of Brittle Solids*, 2nd Ed., Cambridge University Press, Cambridge, U.K., 1993.
4. W. Hirst and M.G.J.W. Howse, "The indentation of materials by wedges," Proc. R. Soc. London, Ser. $\underline{A311}$, 1969, pp. 429–444.
5. M.C. Shaw and D.J. DeSalvo, "A new approach to plasticity and its application to blunt two dimension indenters," J. Eng. Ind. Trans. ASME, $\underline{92}$, 1970, pp. 469–479.
6. G.M. Pharr, W.C. Oliver and F.R. Brotzen, "On the generality of the relationship among contact stiffness, contact area, and the elastic modulus during indentation," J. Mater. Res. $\underline{7}$ 3, 1992, pp. 613–617.
7. M.F. Doerner and W.D. Nix, "A method for interpreting the data from depth-sensing indentation instruments," J. Mater. Res. $\underline{1}$ 4, 1986, pp. 601–609.
8. W.C. Oliver and G.M. Pharr, "An improved technique for determining hardness and elastic modulus using load and displacement sensing indentation experiments," J. Mater. Res., $\underline{7}$ 4, 1992, pp. 1564–1583.
9. C-M. Cheng and Y-T. Cheng, "Further analysis of indentation loading curves: effects of tip rounding on mechanical property measurements," J. Mater. Res. $\underline{13}$ 4, 1998, pp. 1059–1064.
10. K.W. McElhaney, J.J. Vlassak, and W.D. Nix, "Determination of indenter tip geometry and indentation contact area for depth-sensing indentation experiments," J. Mater. Res. $\underline{13}$ 5, 1998, pp. 1300–1306.
11. A. Bolshavok and G.M. Pharr, "Influences of pileup on the measurement of mechanical properties by load and depth sensing indentation techniques," J. Mater. Res. $\underline{13}$ 4, 1998, pp. 1049–1058.
12. C-M. Cheng and Y-T. Cheng, "Effects of 'sinking in' and 'piling up' on estimating the contact area under load in indentation," Philos. Mag. Lett. $\underline{78}$ 2, 1998, pp. 115–120.
13. C-M. Cheng and Y-T. Cheng, "On the initial unloading slope in indentation of elastic-plastic solids by an indenter with an axissymmetric smooth profile," App. Phys. Lett. $\underline{71}$ 18, 1997, pp. 2623–2625.
14. J.S. Field and M.V. Swain, "A simple predictive model for spherical indentation," J. Mater. Res. $\underline{8}$ 2, 1993, pp. 297–306.
15. A.C. Fischer-Cripps, "The Hertzian contact surface," J. Mater. Sci. $\underline{34}$, 1999, pp. 129–137.
16. J.C. Jaeger and N.G.W. Cook, *Fundamentals of Rock Mechanics*, Chapman and Hall, London, 1971.
17. C.A. Coulomb, "Sur une application des règles de Maximis et Minimis a quelques problèmes de statique relatifs à l'Architecture," Acad. R. Sc. Mem. Math. Phys. par divers savans, $\underline{7}$, 1773, pp. 343–382.

Chapter 11
Indentation Test Methods

11.1 Introduction

In previous chapters, we looked in detail at various theoretical aspects of elastic and elastic-plastic indentation. It is now appropriate to discuss how one may undertake this type of indentation testing for particular applications. Hence, we now turn our attention to more practical matters and investigate:

i. Bonded-interface technique
ii. Indentation stress-strain curves
iii. Compliance curves
iv. Strength testing
v. Hardness testing

The information provided here will enable the practitioner to undertake indentation testing that is appropriate to the material and will yield the desired information.

11.2 Bonded-Interface Technique

The bonded-interface specimen technique[1,2] provides a simple and effective way of assessing the subsurface damage arising from indentation loading. The specimens are prepared first by polishing the faces of two rectangular half-specimens and then bonding them together with adhesive. The prospective indentation test surface, perpendicular to the bonded interface, is then polished. As shown in Fig. 11.2.1, a sequence of indentations, centered over the interface, is made on the test surface. A light clamping force is applied during testing in a direction normal to the load to assist the adhesive in keeping the two faces of the specimen in intimate contact during the application of load. After indentation, the specimens are immersed in a solvent to dissolve the adhesive and to separate the specimens into their two halves. Optical microscopy using Nomarski interference contrast can then be used to examine the test surfaces and the subsurface damage and fracture patterns.

The experiment is best performed using a mechanical testing machine which can apply load to the indenter through a cross-head member. Due to the localized nature of the indentation stress field, a relatively large number of indentations may be made on one prepared specimen. It is important that the indentations are centered over the interface.

Estimation of the prospective indentation site by eye is not recommended. In a typical test, the center of the indenter and the interface need to be aligned to within a few microns, and it is recommended that the specimen be mounted on a calibrated stage. Small indentations may then be made on each side of the interface at an arbitrary position. The distance between these indentations and the interface may then be measured with an optical microscope. Calibration for the stage vernier adjustments may thus be found by a geometrical analysis leading to a series of vernier readings which permit indentations to be performed very rapidly and with great precision.

The results of a bonded interface experiment are best viewed with an optical microscope fitted with Nomarksi interference filters. Metallurgical microscopes are also suitable. The indented specimen is placed in a solvent to dissolve the glue, and the two halves are thus separated. After washing in solvent, both top and section views of the indentation can be viewed. The Nomarksi filters highlight very small surface irregularities. The specimen surfaces must not be polished after indentation because this will remove the surface detail, which contains the damage information.

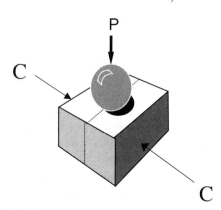

Fig. 11.2.1 Bonded interface schematic. Specimen top and section faces are polished to 1µm. Faces are glued together. Light clamping C is applied to outer faces to assist in preventing fracture of the specimen. Load P is applied to indenter placed on interface. Solvent used to dissolve bond. Top and section faces viewed with Nomarski interference microscope.

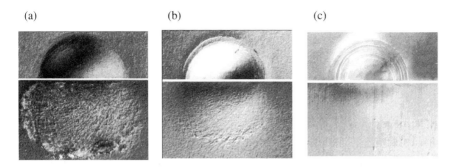

Fig 11.2.2 Bonded interface results. Accumulated damage in both surface (top) and section (bottom) views for a series of coarse-grained glass-ceramics. (a) relatively coarse-grained micaceous glass-ceramic; (b) the same composition as (a) but with a smaller grain size; (c) relatively fine grain size. Results show a transition from classically ductile behavior in (a) to a brittle response (c). The beginnings of a Hertzian cone crack are discernible in the section views of (c) (after reference 3).

Figure 11.2.2 shows accumulated damage in both surface (top) and section (bottom) views for a series of coarse-grained glass-ceramics using this technique. The materials tested here had different microstructural characteristics, and this is manifested in the section views shown. In Fig. 11.2.2, (a) shows the results for a relatively coarse-grained micaceous glass-ceramic, (b) shows the results for a glass-ceramic of the same composition as (a) but with a smaller grain size, and (c) shows the results for the same glass-ceramic with a relatively fine grain size. Note that these results show a transition from classically ductile behavior in (a) to a brittle response (c). The beginnings of a Hertzian cone crack are discernible in the section views of (c).

11.3 Indentation Stress-Strain Response

11.3.1 Theoretical

Valuable information about the elastic and plastic properties of a material can be obtained when the indentation stress, defined as the mean contact pressure p_m, is plotted against the indentation strain, equal to the contact area radius a divided by the indenter radius R for a range of indenter loads and sizes. For a linearly elastic response, a straight-line relationship for these normalized quantities is predicted from the Hertz relation:

$$p_m = \left(\frac{3E}{4\pi k}\right)\frac{a}{R} \qquad (11.3.1a)$$

In Eq. 11.3.1a, k is a dimensionless constant given by:

$$k = 9/16 \, [(1-v^2) + (1-v'^2)E/E'] \qquad (11.3.1b)$$

where v and v' are Poisson's ratio, and E and E' are Young's modulus of the specimen and the indenter, respectively.

Equation 11.3.1a assumes linear elasticity and makes no prediction about the onset of nonlinear behavior followed by plastic yielding within the specimen when the indenter load is sufficiently high. In conventional compression tests, plastic deformation generally does not occur in brittle materials at normal ambient temperatures and pressures due to tensile induced fractures which inevitably occur before the yield stress of the material is reached. However, in tests where there is a significant confining pressure, brittle fracture is suppressed in favor of shear faulting and plastic flow. This phenomenon is familiar to workers in the rock mechanics field[4]. In an indentation stress field, the stresses in the compressive zone beneath the indenter can be made sufficiently high to induce plastic deformation, even in brittle materials.

For the fully elastic case, the principal shear stress distribution beneath a spherical indenter can be readily determined (see Chapter 5) and the maximum shear stress has a value of about $0.47 p_m$ and occurs at a depth in the specimen of about 0.5a beneath the indenter. Tabor[5] uses both the von Mises and Tresca stress criteria to show that plastic deformation beneath a spherical indenter with increasing load can be expected to occur first upon increasing the indenter load when:

$$\begin{aligned} 0.47 p_m &= 0.5 \sigma_y \\ p_m &\approx 1.1 \sigma_y \end{aligned} \qquad (11.3.1c)$$

As the load on the indenter is increased further, the amount of plastic deformation also increases. The mean contact pressure p_m also increases with increasing load. At high values of indentation strain, the response of the material may be predicted using the various hardness theories described in Chapter 9. Experiments[5] show that for metals where the indenter load is such that p_m is about three times the yield stress σ_y, no increase in p_m occurs with increasing indenter load. At this point, the material in the vicinity of the indenter can be regarded as being in a fully plastic state.

11.3.2 Experimental method

Indentation stresses and strains can be measured by recording the indenter loads and corresponding contact diameters of the residual impressions in the surface of gold coated, polished specimens for a range of loads and indenter radii. The specimens should be indented after the deposition of a very thin film of gold, which may be applied using an ordinary sputter coater. The gold film makes the contact diameter easier to distinguish from the unindented surface when the specimen is viewed through an optical microscope. It is important not to make the gold coating too thick, as the measured contact diameter may then be over-estimated.

Figure 11.3.1 shows a worksheet that may be used for an indentation stress-strain experiment. The first two columns indicate a range of indentation strains and indentation stresses calculated using Hertzian theory. The body of the worksheet contains spaces for recording experimentally measured indenter loads and contact diameters for a range of indenter radii. The column on the left of each data entry area shows the load required to give the indicated indentation stress and strain as calculated using Hertzian elastic theory.

In practice, an indentation stress-strain curve of reasonable range cannot be obtained with a single indenter because most testing machines are limited in their load measuring capability. It should be noted that a particular value of indentation stress and strain may be obtained with different indenter sizes at different loads. Some overlap between the range of stresses and strains with different indenters gives a convenient check of the validity of the experimental procedure.

Many materials can be considered elastic-plastic where the transition from elastic to plastic occurs very suddenly in brittle materials. In the ideal case, the indentation stress-strain relationship is expected to show an initial straight-line response, as given by Eq. 11.3.1, followed by a decrease in slope until the indentation stress approaches that corresponding approximately to the hardness value H.

Figure 11.3.2 shows the results for a coarse-grained micaceous glass-ceramic (see also Fig. 11.2.2). The solid line shows the Hertzian elastic response as calculated using Eq. 11.3.1. Finite-element results and experimental measurements for WC spheres of radius R = 0.79, 1.59, 1.98, 3.18, 4.76 mm are also shown together with hardness computed from the projected area of indentation with a Vickers diamond pyramid.

For a brittle material, one cannot expect to obtain the full stress-strain response using the experimental procedure described here because of the inevitable presence of conical fractures which would occur at high indenter loads. Indeed, one would be very fortunate to obtain indentation stresses and strains in the nonlinear region for brittle materials without the presence of a significant number of conical fractures and perhaps bulk specimen failure. An indentation stress-strain response for a ductile material is readily obtained using this test procedure. For a brittle material, only a relatively narrow range of readings can be obtained with a spherical indenter, and attempts to obtain data at higher strains will probably result in conical fractures and specimen failure.

Indentation Stress-Strain (Experiment)

Page #

Elastic properties

	I	Specimen	
Material	WC		
E	614	70	GPa
υ	0.22	0.26	
k	1.12 (as per M&M)		
k	0.59 (as per Frank&Lawn)		
Step	0.02 mm		

This worksheet provides an estimate of the indenter loads required for different ball radii which gives a selected range of contact pressures. The smallest radii provide high contact pressures at a moderate load. Larger radii cannot be used for a load cell maximum of 5kN. Select a range of loads of about 400-1200N within each ball radius column and allow one or two overlapping load/radius pair in each which gives the same contact pressure. Data in grey boxes may be changed by the user.

Suggested indenter loads (N)

$$P_m = \frac{4}{3} E \frac{1}{(1-\upsilon^2) k\pi} \frac{a}{R} \qquad P_{suggestd} = p_m \pi \left(\frac{a}{R} R\right)^2$$

x axis desired a/R	Mean pressure GPa	R 0.79 Suggested	Actual	Contact diameter	R 1.19 Suggested	Actual	Contact diameter	R 1.59 Suggested	Actual	Contact diameter
0.02	0.57									
0.04	1.14							127		
0.06	1.71							227		
0.08	2.28					116		392		
0.10	2.85				127			622		
0.12	3.43				219			929		
0.14	4.00	154			348			1322		
0.16	4.57	229			520			1813		
0.18	5.14	326			741			2414		
0.20	5.71	448			1016			3134		
0.22	6.28	596			1352					
0.24	6.85	774			1755					
0.26	7.42	984			2232					
0.28	7.99	1228			2787					
0.30	8.56	1511								
0.32	9.13	1834								
0.34	9.70	2199								

a/R	GPa	R 1.98 Suggested	Actual	Contact diameter	R 3.18 Suggested	Actual	Contact diameter	R 4.76 Suggested	Actual	Contact diameter
0.02	0.57									
0.04	1.14				58			130		
0.06	1.71	76			196			439		
0.08	2.28	180			464			1040		
0.10	2.85	352			907			2032		
0.12	3.43	607			1567					
0.14	4.00	965			2488					
0.16	4.57	1440								
0.18	5.14	2050								
0.20	5.71	2812								

		R 6.35 Suggested	Actual	Contact diameter	R 7.94 Suggested	Actual	Contact diameter	R 9.53 Suggested	Actual	Contact diameter
0.025	0.71	56			88			127		
0.030	0.86	98			153			220		
0.035	1.00	155			242			349		
0.040	1.14	231			362			521		
0.045	1.28	329			515			742		
0.050	1.43	452			707			1018		
0.055	1.57	602			941					

Fig. 11.3.1 Worksheet for indentation stress-strain experiment. Note that the indenter sizes and loads have been selected to give some overlap in indentation strains as the indenter is changed. The important parameter is the quantity a/R, the indentation strain. For a particular test, it may not be possible to use a single indenter to cover the desired range of indentation strain. However, the load and indenter radius in different combinations may permit a wide range of indentation strains to be measured with a readily available apparatus.

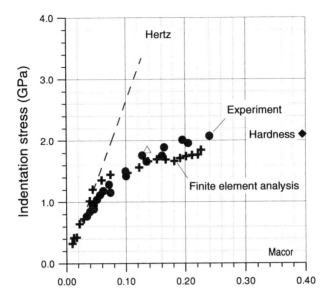

Fig. 11.3.2 Indentation stress strain results for a coarse-grained micaceous glass-ceramic. Solid line shows Hertzian elastic response. (+) Finite-element results, (•) experimental measurements for WC spheres of radius R = 0.79, 1.59, 1.98, 3.18, 4.76 mm, and hardness from projected area of indentation with Vickers diamond pyramid are shown (data after reference 3).

11.4 Compliance Curves

Compliance curves are obtained by measuring the load point displacement (see Chapter 6). Typically, a polished specimen is mounted on the horizontal platen of a universal testing machine. Load is applied to an indenter by causing the crosshead of the testing machine to move downward at a constant rate of displacement with time. A clip gauge is attached so as to measure the load-point displacement as shown in Fig. 11.4.1.

The output signal from the clip gauge is often interfaced to a computer system that records displacement at regular time intervals during the application of load. A fully elastic response, for a spherical indenter, is given by:

$$\delta = \frac{a^2}{R} \tag{11.4a}$$

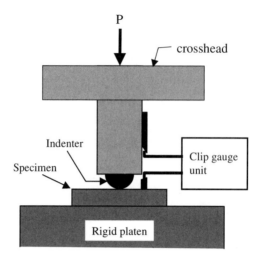

Fig. 11.4.1 Compliance testing schematic. Specimen is mounted on the horizontal platen of a universal testing machine. Load is applied to an indenter by causing the crosshead of the testing machine to move downward at a constant rate of displacement with time. A clip gauge is attached to measure the load-point displacement. The output from the clip gauge is often interfaced to a computer system that records displacement at regular time intervals during the application of load (after reference 6).

In order to obtain meaningful results from a compliance test, it is necessary to make certain corrections to the clip-gauge output, depending on the nature of the experimental apparatus. For example, in Fig. 11.4.1, the clip gauge is mounted between "fixed" points on the crosshead post and the indenter surface. The slip gauge thus measures both the indentation and the longitudinal strain arising from the compression of the post. All that is actually required is the displacement of the crosshead arising from indentation of the specimen. Further, depending on the geometry and mounting arrangements of the indenter, the indenter may indent into the crosshead post in addition to the specimen under test. For this reason, it is best to use an indenter with a relatively large area upper surface. For example, with spherical indenters, a cup or half-sphere should be used.

Figure 11.4.2 shows the results obtained on specimens of glass and a coarse-grained glass-ceramic material. The glass displays a characteristic brittle response, and the displacements are in reasonable agreement with those calculated using Eq. 11.4a. The glass-ceramic undergoes shear-driven "plasticity" (see Fig. 11.2.2) and thus displays a considerable deviation from the elastic response. The area under these curves is an indication of the energy associated with the indentation (see Fig. 10.6.1). Note that the unloading curve meets the x axis at a depth corresponding to that of the residual impression in the specimen surface.

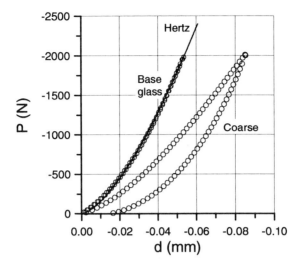

Fig. 11.4.2 Compliance curves, loading, and unloading, for glass (elastic) and glass-ceramic (elastic-plastic) specimen materials. Specimens were loaded with a spherical indenter R = 3.18 mm at a constant rate. Displacement was recorded at regular intervals. Solid line indicates Hertzian elastic response calculated using Eq. 11.4a. Data are shown for experiments performed on glass and a coarse-grained glass-ceramic. Note the residual impression upon full unload for the coarse-grained ceramic, indicating plastic deformation (data after reference 6).

The discussion thus far applies to indentations on an engineering scale, that is, where the dimensions of the indenter and specimen are measured in millimeters and loads in N or even kN. In many cases, useful material properties and physical insights on damage on a microscopic scale can be obtained submicron indentation systems. These machines use micron-size indenters and mN loadings to produce extremely shallow indentations in test materials. Such machines are particularly suited for measuring the mechanical properties of thin films. Because of this small scale, these instruments are typically computer controlled, with the test specimen and loading mechanism located in a protective cabinet.

11.5 Inert Strength

Bending strength tests provide a quantitative measure of damage caused by indentation with a sharp or blunt indenter. Theoretical analysis shows that for brittle materials with a constant value of toughness, the following relation holds for a well-developed cone crack in a previously indented specimen[7]:

Chapter 11. Indentation Test Methods

$$\sigma_I = \left(\frac{T_o^4}{\psi^3 \chi P}\right)^{1/3} \quad (11.5.1)$$

In this equation, σ_I—the "inert strength"—is the macroscopic tensile stress applied to the specimen during bending. T_o is the toughness, and ψ and χ are constants found from theoretical analysis and experimental calibration, respectively. P is the indenter load used to indent the specimen on the prospective tensile side prior to bending. Equation 11.5.1 shows that an ideal elastic response, with a constant value of T_o, gives σ_I proportional to $P^{-1/3}$. This relationship applies to well-developed cone cracks, which generally occur in classical brittle materials. The relationship between P and σ_I for material showing accumulated subsurface damage is not currently defined.

In a typical experiment, bars of the specimen material are prepared and the prospective test faces polished to a 3 μm finish. The edges of each bar are chamfered to minimize edge failures during the test. A single indentation is made on the polished face of each specimen using, say, a 3.18 mm WC sphere. Some specimens are left unindented to measure the "natural" strength of the material. The specimens are then loaded at a rate of 1000 N/sec in four-point bending so that the polished, indented surface is placed in tension (see Fig. 11.5.1).

The tensile stress σ_I on the test surface at a measured failure load P is calculated from:

$$\sigma_I = \frac{6P}{2t^2 w} \frac{(L-l)}{2} \quad (11.5.2)$$

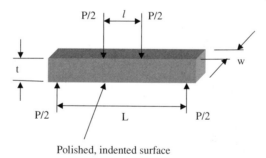

Polished, indented surface

Fig. 11.5.1 Schematic of strength experiment using prismatic bar specimens. Specimens are polished and edges chamfered. A single indentation is made on the polished surface. The bar is then put into 4 point bending with the indented surface being placed in tension as shown.

where t is the thickness and w the width of the specimen. L is the outer span and *l* the inner span, as shown in Fig. 11.5.1. The load P is that indicated by the testing machine at specimen fracture. This may not always be easy to determine, and the use of a calibrated piezoelectric force transducer may be required. Figure 11.5.2 shows the results of such a test on a glass-ceramic material.

In Fig. 11.5.2, the shaded box on the left indicates the strength of the specimens that failed at a location away from where the indentation was made (i.e., from "natural" flaws). In this figure, results for both the base-glass state and the fired, crystallized material are shown. In contrast to the base-glass state of the material, the strength data shown for the crystallized material show that the tensile strength is not significantly affected by the presence of the subsurface accumulated damage (see Fig. 11.2.2) beneath the indentation site, although an overall decrease in strength with increasing indenter load is indicated. In the case of the base-glass, a cone crack forms above a critical indenter load, the magnitude of which depends on the specimen surface condition and the radius of the indenter.

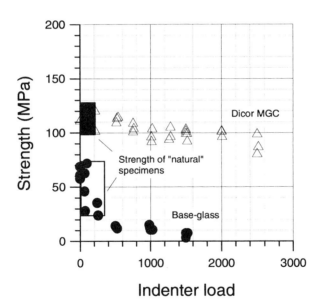

Fig. 11.5.2 Strength of glass-ceramic and base-glass after indentation with a WC spherical indenter of radius R = 3.18 mm. Shaded area indicates range of strengths for samples that failed from a flaw other than that due to the indentation. Also included in the shaded areas are the strengths of a small number of unindented samples. Each data point represents a single specimen. Solid curves are empirical best fits to the data. For the base-glass, the critical indenter load for the formation of a cone crack was ≈500 N.

In the present case, the critical load for formation of a cone crack is ≈500 N for an indenter radius R = 3.18 mm. The results shown in Fig. 11.5.2 indicate that the tensile strength of the base-glass is significantly affected by the presence of a conical crack, and the strength is reduced as the size of the crack is made larger (increasing indenter load). Note the increased variability of the strength (height of shaded area) of the unindented specimens of the base-glass compared to the crystallized glass-ceramic material.

11.6 Hardness Testing

11.6.1 Vickers hardness

The Vickers hardness number is one of the most widely used measures of hardness in engineering and science. In a typical hardness tester, the diamond indenter is mounted on a sliding post brought to bear on the specimen, which is mounted on the flat movable platen. The indenter and mechanism can then be swung to the side and a calibrated optical microscope positioned over the indentation to measure the dimensions of the residual impression. Figure 11.6.1 shows typical shapes of indentations made with a Vickers indenter.

The Vickers diamond indenter takes the form of a square pyramid with opposite faces at an angle of 136° (edges at 148°). The Vickers diamond hardness, VDH, is calculated using the indenter load and the *actual* surface area of the impression. The resulting quantity is usually expressed in kg/mm². The area of the base of the pyramid, at a plane in line with the surface of the specimen, is equal to 0.927 times the surface area of the faces that actually contact the specimen. The mean contact pressure p_m is given by the load divided by the projected area of the impression. Thus, the Vickers hardness number is lower than the mean contact pressure by ≈7%. In many cases, scientists prefer to use the projected area for determining hardness because this gives the mean contact pressure—a value of some physical significance—while also providing a comparative measure of hardness. The hardness calculated using the actual area of contact does not have any physical significance and can only be used as a comparative measure of hardness.

The Vickers diamond hardness is found from:

$$\text{VDH} = \frac{2P}{d^2} \sin \frac{136°}{2}$$
$$= 1.854 \frac{P}{d^2}$$
(11.6.1a)

with d the length of the diagonal as measured from corner to corner on the residual impression. The projected area of contact can be readily calculated from a measurement of the diagonal and is equal to:

$$A_p = \frac{d^2}{2} \tag{11.6.1b}$$

The ratio between the length of the diagonal d and the depth of the impression h beneath the contact is 7.006, and thus the projected area A_p, in terms of the depth h, is equal to:

$$A_p = 24.504h^2 \tag{11.6.1c}$$

The residual impression in the surface of a specimen made from a Vickers diamond indenter may not be perfectly square. Depending on the material, the sides may be slightly curved to give either a pin-cushion appearance (sinking in—annealed materials) or a barrel-shaped outline (piling up—work-hardened materials) as shown in Fig. 11.6.1. It is a matter of individual judgment whether the curved sides of the impression should be taken into consideration when determining the contact area. The formal definition of the Vickers hardness number[8] involves the use of the mean value of the two diagonals, regardless of the shape of the sides of the impression.

Various experimental factors affect the value of VDH as calculated using Eq. 11.6.1a. Vickers hardness data are usually quoted together with the load used and the loading time. The loading time, which is normally 10–15 seconds, is that at which *full* load is applied. The load should be applied and removed smoothly. Although the load rate is not specified in the ASTM Standards[8], McColm[9] claims that a load rate of less than 250 µm s^{-1} is required for low load applications in order to avoid the calculation of artificially low hardness values.

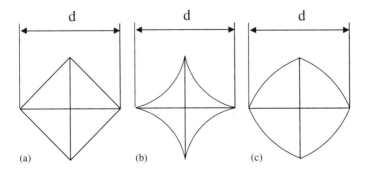

Fig. 11.6.1 Residual impression made using a Vickers diamond pyramid indenter. (a) Normal impression, (b) Sinking in, (c) Piling up.

11.6.2 Berkovich indenter

Microhardness testing on a very small scale is conveniently carried out using a Berkovich three-sided pyramidal indenter, since the facets of the pyramid may be constructed to meet at a single point rather than a line, which usually results at the apex of a four-sided pyramidal indenter. The Berkovich indenter is thus very useful for the investigation of the mechanical properties of thin films such as optical coatings, paint, and hard coatings on machine tools. However, due to the small scale of the impressions, measurements of the radius of the circle of contact are extremely difficult, often requiring the use of expensive electron microscopes. For this reason, much attention has been paid to the estimation of the radius of circle of contact from the depths of the fully loaded or unloaded impressions, which are more easily measured.

The included angle of the Berkovich indenter (65.3°) gives the same projected area for the same depth of penetration as the Vickers indenter (angle 68°). If d is the length of one side of the triangular impression, then the projected area of contact is given by:

$$A_p = 0.4330 1 d^2 \qquad (11.6.2a)$$

The depth of the impression beneath the contact is:

$$h = 0.1327 d \qquad (11.6.2b)$$

which gives a projected area in terms of the depth of:

$$A_p = 24.59 h^2 \qquad (11.6.2c)$$

which is comparable with Eq. 11.6.1c.

Note that the hardness number associated with all these indenters refers to the area of contact of the impression measured at the point at which the indenter meets the surface of the specimen in the fully loaded condition. However, the vast majority of such measurements are made using the size of the residual impression (i.e., fully unloaded condition). Experiments show that while the depth of penetration beneath the contact may vary appreciably due to elastic recovery of the specimen when load is removed, there is little difference between the size of the radius or diagonal measurements of the contact area.

11.6.3 Knoop hardness

The Knoop indenter was invented by F. Knoop at the National Bureau of Standards (NBS) (now the National Institute of Standards & Technology) in the United States. As indicated in Fig. 11.6.3, the angles for the opposite faces of a Knoop indenter are 172°30′ and 130°. The Knoop indenter is particularly useful for the study of highly brittle materials due to the smaller depth of penetration for a given indenter load. Further, due to the unequal lengths of the diagonals, it is also very useful for investigating anisotropy of the surface of the specimen.

As shown in Fig. 11.6.3, the length d of the longer diagonal is used to determine the projected area of the impression. The Knoop hardness number calculated from[*]:

$$\text{KHN} = \frac{2P}{d^2 \left[\cot\frac{172.5}{2} \tan\frac{130}{2} \right]} \qquad (11.6.3a)$$

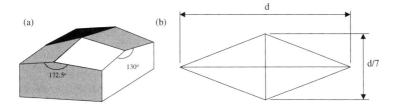

Fig. 11.6.3 (a) Geometry of Knoop indenter, (b) Residual impression.

1. S.R. Williams, *Hardness and hardness measurements*, American Society for Metals, Cleveland, Ohio, USA, 1942.
2. T.O. Mulhearn, "The deformation of metals by Vickers-type pyramidal indenters," J. Mech. Phys. Solids, 7, 1959, pp. 85–96.
3. A.C. Fischer-Cripps, "A Partitioned-problem approach to microstructural modelling of a glass-ceramic," J. Am. Ceram. Soc. 1999 in press.
4. K.T. Nihei, L.R. Myer, J.M. Kemeny, Z.Liu and N.G.W. Cook, "Effects of heterogeneity and friction on the deformation and strength of rock," in *Fracture and damage in quasi-brittle structures: Experiment, Modelling and Computer Analysis*, Eds. Z. P. Bazant, Z. Bittnar, M. Jirasek and J. Mazars, E&F Spon, London, 1994.
5. D. Tabor, *The Hardness of Metals*, Clarendon Press, Oxford, 1951.
6. A.C. Fischer-Cripps, "The Hertzian contact surface," J. Mater. Sci., 34, 1999, pp. 129–137.
7. B.R. Lawn, *Fracture of Brittle Solids* 2nd. Ed. Cambridge University Press, Cambridge, U.K., 1993.
8. ASTM "Standard test method for Vickers hardness of metallic materials," E92, in *Annual Book of ASTM Standards, Section 3, Volume 3.01, Metals-Mechanical Testing, Elevated and Low Temperature Tests*, ASTM, Philadelphia, PA, Edited by R.A. Storer, 1993.
9. I.J. McColm, *Ceramic Hardness*, Plenum Press, New York, 1990.

---

[*] It should be noted that Eq. 11.6.3a as it appears in reference 9 is misprinted.

Appendix 1
Submicron Indentation Test Analysis

A1.1 Introduction

In conventional indentation tests, the area of contact between the indenter and the specimen at maximum load is usually calculated from the diameter or size of the residual impression after the load has been removed. The size of the residual impression is usually considered to be identical to the contact area at full load, although the depth of penetration may of course be significantly reduced by elastic recovery. Direct imaging of residual impressions made in the submicron regime are usually only possible using inconvenient means and for this reason, it is usual to measure the load and depth of penetration directly during loading and unloading of the indenter. These measurements are then used to determine the projected area of contact for the purpose of calculating hardness and elastic modulus and these methods have been described in Chapter 10. In practice, various errors are associated with this procedure. The most serious of these errors manifests themselves as offsets to the depth measurements and arise from:

1. the initial penetration depth
2. the compliance of the loading system
3. the nonideal shape of the indenter

In this appendix, further details of the analysis procedure the origin of these errors is discussed and the methods used to correct experimental data for them described.

A1.2 Initial Penetration Depth

Indentation tests in the submicron regime are usually performed by bringing the indenter into contact with the specimen with a controlled force and then measuring the resulting depth of penetration. The penetration depth is ideally measured from the specimen free surface. However, in practice, the indenter must make contact with the specimen surface before the depth measurements can be taken. That is, it is not possible to establish a datum for the depth measurements before an initial contact with the surface is made. This "initial contact" is usually made as small as possible and is often set to the smallest measurable force of the instrument. An initial contact force on the order of 0.01 mN is usually achievable. No matter how small the initial contact force is made, there is a corresponding penetration of the indenter beneath the specimen surface. Thus, all subsequent depth measurements taken from this datum will be in error by this small initial penetration depth. The initial penetration depth h_i has to be added to all depth measurements h to correct for this initial displacement.

How might this initial penetration depth be estimated? First, we might assume that the first few loading points result in an elastic deformation of the specimen. This is certainly reasonable for a spherical indenter, but also may be applicable for a pointed indenter since in practice, the tip of a pointed indenter is not perfectly sharp. One may then fit a suitable least squares regression curve to these initial load-displacement data points and extrapolate the fitted curve back to zero load. Care should be taken to only use the elastic response for the fitting since if too many data points at an increasing load are chosen, then the fit will be affected by plastic deformation in the specimen. The fitting and extrapolation should be done for a number of different sets of experimental data collected under the same conditions and an average h_i taken and subsequently applied to the experimental data.

A second method is more elaborate and seeks to account for the actual expected shape of the loading curve. As before, we assume that the initial contact, and the response for a few load steps afterward, is elastic. In general, the penetration depth for elastic contact for axis-symmetric indenters is such that:

$$h \propto P^n \tag{A1.2a}$$

where n = 2/3 for a spherical indenter, n = 1 for a cylindrical flat punch indenter, and n = 1/2 for a conical indenter. At a certain force P_i, the initial contact load, there is an initial depth h_i. During the initial loading of the specimen—where the response is elastic—the instrument measures P and h_t where the total penetration $h_t' = h_t + h_i$. Thus,

$$h_t + h_i \propto P^n \tag{A1.2b}$$

and therefore

$$\begin{aligned} h_t &\propto P^n - h_i \\ &= k\left(P^n - P_i^n\right) \end{aligned} \tag{A1.2c}$$

where k is a constant with a value that depends on the shape of the indenter. For an initial elastic response, say, for the first 5 or so data points in a typical test, we have a series of values for h_t and P, and also a value for P_i. The terms n and k are the unknowns. Once values for n and k are found, the initial penetration h_i can be found from:

$$h_i = kP_i^n \tag{A1.2d}$$

From Eq. A1.2c, a plot of h_t vs $(P^n - P_i^n)$ should be linear with a slope k. The easiest way to adjust the variable n and k for a linear response is to plot the logarithm of both sides to obtain a slope equal to unity. Thus:

$$\log h_t = \log k + m \log(P^n - P_i^n) \qquad (A1.2c)$$

where m =1. A plot of log h_t vs log $(P^n-P_i^n)$ should have a slope of 1 if n and k are chosen correctly. Note that the first data point of the experimental data with $h_m=0$ cannot be included since log 0 is not defined. The procedure is thus to start with n=2/3. The initial contact force P_i and measured values of P and h_t are plotted according to Eq. A1.2c and the value of n until the slope m is as close to unity as possible. An additional fitting may be required, since there is in practice an error associated with the minimum contact force. Thus, it may be beneficial to adjust P_i for a minimum value of m and then adjust n.

The resulting intercept provides a value for k. A value for h_i can then be determined from Eq. A1.2d. The corrected depth h_t' is thus:

$$h_t' = h_t + h_i \qquad (A1.2d)$$

Care must be taken with the choice of the number of initial data points to be used in the analysis. It is assumed that the material response is elastic and thus not too many points should be used, since then the data used for the regression will contain that arising from plastic deformation. Further, estimates of h_i should be obtained on a series of tests performed under the same conditions but with different maximum contact loads to ensure repeatability.

A1.3 Instrument Compliance

Doerner and Nix[1] observed that for tests with a Berkovich indenter, the *initial* unloading curve is linear for a wide range of test materials, which implied a response similar to that observed for the elastic unloading of a cylindrical punch. These workers then applied cylindrical punch equations to the initial part of an unloading curve to determine the size of the contact from depth measurements. Oliver and Pharr[2] extended this treatment to account for observed changes in the area of contact during initial unloading by treating the elastic unloading in terms of both a cone and a paraboloid of revolution. The total depth of penetration h_t, less the distance from the contact to the specimen free surface h_s, is the plastic depth h_p. A correction factor is applied to h_s to determine the plastic depth h_p and hence the hardness and modulus. The constant K equals 1 for cylindrical contact, K equals 0.72 for a cone, and 0.75 for a paraboloid.

$$h_s = KP_{max} \frac{dh}{dP} \qquad (A1.3a)$$

$$h_p = h_t - h_s$$

In indentation tests, the penetration beneath the specimen free surface (corrected for the initial contact depth) h_t' is measured. In the case of a Berkovich indenter, the slope of the initial unloading curve (dP/dh' at P_{max}) may be used to determine h_s and hence h_p. The elastic modulus is found from:

$$\frac{dP}{dh} = \frac{2h_p E}{1-v^2} \sqrt{\frac{24.5}{\pi}} \tag{A1.3b}$$

Equation A1.3b shows that the initial slope of the unloading curve dP/dh is proportional to E for a given plastic penetration h_p. This initial slope is sometimes referred to as the "unloading stiffness" and the inverse, dh/dP, the specimen "compliance."

However, the depth measuring system registers the depth of penetration of the indenter into the specimen and also any displacements of the loading column arising from reaction forces during loading. These displacements are proportional to the load. Thus, the unloading stiffness has contributions from both the elastic responses of the specimen and the instrument. Rearranging Eq. A1.3b, and including the compliance of the instrument C_f, we have:

$$\frac{dh}{dP} = \left[\sqrt{\frac{\pi}{24.5}} \frac{1}{E^*}\right] \frac{1}{2h_p} + C_f \tag{A1.3c}$$

where we have replaced $E/(1-v^2)$ with the composite modulus E^* for generality. If a series of tests is run at different maximum loads (and hence different values of h_p), a plot of dh/dP vs $1/h_p$ should be a straight line with a slope proportional to $1/E^*$ and an intercept that gives the compliance of the instrument C_f directly. The compliance of the instrument may thus be subtracted from experimental values of dh/dP before calculating E^*. Alternatively, since the displacements arising from instrument compliance are proportional to the load, a correction may be made to the indentation depths h_t' (already corrected for initial contact) to give a further corrected depth h_t'' according to:

$$h_t'' = h_t' - C_f P \tag{A1.3d}$$

For a spherical indenter, measured depths h_t' may be corrected in the same way to account for the depth contribution arising from instrument compliance.

Another method of determining C_f is to test a specimen of known compliance. The measured compliance dh/dP may be compared with the expected or known compliance, with the difference between the two being an estimation of C_f.

Two corrections can thus far be made to the measured depths. The first correction accounts for the initial displacement at the initial contact loading, and the second accounts for the compliance of the instrument.

A1.4 Indenter Shape Correction

The equations presented so far have assumed that the geometry of the indenter is ideal. This is seldom the case in practice, particularly with indenters used for submicron testing. During the course of the investigation into the nature of the initial unloading for a Berkovich indenter, Pharr, Oliver, and Brotzen[3] showed that the relationship between the contact area, the elastic modulus, and the unloading stiffness for a cylindrical punch indenter applied to all axis-symmetric indenters of a smooth profile:

$$\frac{dP}{dh} = 2E^* \frac{\sqrt{A}}{\sqrt{\pi}} \qquad (A1.4a)$$

For example, for a Vickers or a Berkovich indenter, the relationship between the projected area A of the indentation and the depth h_p beneath the contact is:

$$A = 24.5 h_p^2 \qquad (A1.4b)$$

Thus:

$$\frac{dP}{dh} = 2h_p E^* \sqrt{\frac{24.5}{\pi}} \qquad (A1.4c)$$

In the case of a spherical indenter, we need not make use of Eq. A1.4a since we have the radius of the circle of contact at h_p given by geometry:

$$a = \left[2Rh_p - h_p^2\right]^{1/2} \qquad (A1.4d)$$

In Eq. A1.4d, R is the radius of the indenter, which for the spherical indenter analysis procedure is assumed to be rigid. The area of indentation can be found from:

$$A = \pi\left[2Rh_p - h_p^2\right] \qquad (A1.4e)$$

where h_p is found from

$$h_p = \frac{h_t + h_r}{2} \qquad (A1.4f)$$

and h_r from

$$h_r = \frac{h_s (P_t/P_s)^{2/3} - h_t}{(P_t/P_s)^{2/3} - 1} \qquad (A1.4g)$$

Equations A1.4b and A1.4e are "area functions" which relate the projected area of contact with the penetration depth. These equations assume that the geometry of the indenter is ideal, a circumstance impossible to achieve in practice.

222   Appendix 1. Submicron Indentation Test Analysis

The most significant departure from nonideal geometry occurs at the tip of pointed indenters, where blunting causes the measured depth of penetration for a given contact area to be less than would be achieved with a perfectly sharp indenter. For spherical indenters, especially those made from diamond, the crystalline structure of the material usually results in a nonspherical profile at the nanometer scale. It is therefore necessary to apply a correction to measured data to account for the nonideal geometry of the indenter used in any practical test. This correction is usually applied as a modified area function used to calculate E and H.

The usual method of determining the modified area function requires that several indentations be made on a test specimen at different maximum depths. The projected area of contact is measured by direct imaging (e.g., using an SEM). The measured area A_m from the SEM is then plotted against the plastic depth h_p determined from the measured depths (corrected for compliance and initial contact). Regression analysis of the appropriate order may then provide an analytical function that gives the actual projected area for a given value of h_p and this area used in Eq. A1.4a to calculate E and H ($=P_{max}/A$).

The shape correction applies to the area function, not the values of measured depth h_t. The indentation testing machine provides values of depth h_t (which may be further corrected for initial penetration and compliance). The purpose of the area function is to provide a means of estimating the actual contact area for this measured penetration for the purposes of calculating H and $E^{\prime *}$.

The disadvantage of the approach above is that direct imaging of a series of indentations is required to obtain the area function calibration curve. This is not always convenient, especially for very low loads, where tip rounding and area correction are most important and where direct imaging is the most difficult to perform. There is an alternate method of obtaining the area function if the elastic properties of the specimen material are known beforehand. The procedure is to perform a series of indentations at varying maximum loads on a standard test specimen whose elastic modulus and Poisson's ratio are known. If E^* is known (embodying the elastic properties of both indenter and specimen), then the actual area of contact at each load, for a Berkovich indenter, is found from:

$$A = \left[\frac{dP}{dh} \frac{\sqrt{\pi}}{2} \frac{1}{\beta E^*} \right]^2 \qquad (A1.4h)$$

---

* Alternatively, one may take the view that we could also determine the effective plastic depth h_p that would result in the actual contact area for an ideal tip. However, this would be misleading since it is only the value of h_p that requires correction, not *all* the values of penetration h_t. If the correction is applied to h_t, then the value of dP/dh is unaffected, but the absolute values of depth no longer represent those actually achieved during the test but rather those that would have been achieved if an ideal indenter had been used. In this case, Eq. A1.4c would be used directly, since the tip correction would have been applied to the depth measurements. This has the effect of translating the load versus depth data to the right on a compliance curve.

where dP/dh is evaluated from the initial unloading. β is a geometrical correction factor which is discussed later. For a spherical indenter, the actual area of contact is:

$$A = \pi \left[\frac{3}{4} \frac{P}{E^*(h_t - h_r)} \right]^2 \qquad (A1.4i)$$

where each test corresponds to a different value of h_p. Values of A and h_p for each test on the reference material provide the data for a shape correction lookup or calibration table.

It is sometimes convenient to express the shape correction as a ratio of the actual area to the ideal area of indentation. For a Berkovich indenter, the actual area A is given by Eq. A1.4h and the ideal area A_i from Eq. A1.4b. The corrected hardness H and modulus are thus found from:

$$H = \frac{P}{A}$$

$$= \frac{P}{24.5 h_p^2} \frac{A_i}{A} \qquad (A1.4j)$$

where A_i/A is a function of h_p. Also,

$$E^* = \frac{dP}{dh} \frac{1}{2h_p} \sqrt{\frac{\pi}{24.5}} \sqrt{\frac{A_i}{A}} \qquad (A1.4k)$$

For a spherical indenter, the corrected hardness and modulus are:

$$H = \frac{P}{\pi(2Rh_p - h_p^2)} \frac{A_i}{A} \qquad (A1.4l)$$

and

$$E^* = \frac{3}{4} \frac{P}{(2Rh_p - h_p^2)^{1/2}(h_t - h_r)} \sqrt{\frac{A_i}{A}} \qquad (A1.4m)$$

In Eq. A1.4m, h_t is measured at load P_t, and h_r can be found from measurements of h_t and h_s and a load P_t and a partial unload P_s according to Eq. A1.4g. It should be remembered that the spherical indenter analysis procedure assumes a rigid indenter and that E^* in Eq. A1.4m contains E and v for the specimen only (see Chapter 10).

In the case of a Berkovich or Vickers indenter, a further correction may be applicable to account for the nonaxis-symmetric nature of the indenter. This correction, termed β, is applied to Eq. A1.4a and is equal to 1.034 for a Berkovich indenter, 1.012 for a Vickers indenter, and 1.00 for a sphere. This factor is a result of finite-element calculations for indentations formed with flat-ended punches of triangular and square cross-sections[4] and seeks to account for the differences in response between axis-symmetric indenters and their symmetric counterparts. Thus, Eq. A1.4a can be written:

$$\frac{dP}{dh} = \frac{2\beta E^* \sqrt{A}}{\sqrt{\pi}} \quad \text{(A1.4n)}$$

As a final note, Eq. A1.3c implicitly assumes an ideal tip (the factor 24.5 is used), but the values of penetration depth used for calculating dP/dh, which is used to determine the area function in Eq. A1.4c, may have already been corrected for compliance (the instrument compliance is effectively subtracted from the measured value of dh/dP). The significance of this is that the compliance correction and the area function determination are not independent. Once an area function has been determined, it may be necessary to recalculate the compliance correction term and then repeat the area function determination until there is little or no change in the computed quantities.

A1.5 Hardness as a Function of Depth

For tests with a Berkovich indenter, the hardness H and the modulus E^* are determined from the slope dP/dh and the plastic depth h_p at maximum load. Once the slope at maximum load is known, then, assuming that the modulus E^* is constant (i.e., independent of depth), the hardness may be computed at any point along the loading curve. At any points P and h_t along the loading curve:

$$h_p = h_t - h_s$$
$$h_s = KP \frac{dh}{dP} \quad \text{(A1.5a)}$$
$$h_p = h_t - KP \frac{dh}{dP}$$

The constant K equals 1 for a hypothetical cylindrical punch unloading, and K=0.75 for a paraboloid of revolution. Now, h_t and P are both provided (after certain corrections) by the indentation testing machine. The question is: "What is the value of dP/dh at any point P?" The only value of dP/dh available thus far

Appendix 1. Submicron Indentation Test Analysis

is that measured at maximum load P_{max}. However, if E^* does not depend on indentation depth, then Eq. A1.4c, applies for *any* load P. Thus:

$$\frac{dP}{dh}\frac{1}{2h_p}\sqrt{\frac{\pi}{24.5}} = \frac{dP}{dh}_{max}\frac{1}{2h_{p\,max}}\sqrt{\frac{\pi}{24.5}}$$

$$\frac{dh}{dP} = \frac{dh}{dP}_{max}\frac{h_{p\,max}}{h_p}$$

(A1.5b)

Equation A1.5b shows that dh/dP at load P is a function of h_p at that point. Note also that we have ignored the differences in area function correction that would apply at the different loads P and P_{max}. Inserting Eq. A1.5b into Eq. A1.5a, we obtain:

$$h_p = h_t - KP\frac{dh}{dP}_{max}\frac{h_{p\,max}}{h_p}$$

(A1.5c)

which is a quadratic equation in h_p and where K is the correction factor equal to 0.75 for a Berkovich indenter. The hardness H at load P can then be found from:

$$H = \frac{P}{24.5 h_p^2}\frac{A_i}{A}$$

(A1.5d)

If the variation of hardness H with depth is expected to be small, then a rough estimate of the quantity dP/dh at load P may be made by substituting Eq. A1.4j into A1.5b. With H a constant, then:

$$\frac{dP}{dh}\frac{\pi}{2}\frac{\sqrt{H}}{\sqrt{P}} = \frac{dP}{dh}_{max}\frac{\pi}{2}\frac{\sqrt{H}}{\sqrt{P_{max}}}$$

$$\frac{dP}{dh} = \frac{dP}{dh}_{max}\frac{\sqrt{P}}{\sqrt{P_{max}}}$$

(A1.5e)

Equation A1.5e has the advantage of not requiring a solution to a quadratic equation but only applies when the H is not a strong function of depth of penetration and that, along with Eq. A1.5c, E^* is a constant with depth.

A value of hardness may be more accurately determined at intermediate depths by a partial unload during the loading sequence. In this way, direct measurements of dP/dh are available from which h_p, and hence H, is obtained from Eqs. A1.5a and A1.5d.

A1.6 Generating Simulated Data

A1.6.1 Berkovich indenter

For an elastic-plastic response, the relationship between the total depth of penetration and the load for indentation with a Berkovich indenter can be found from:

$$h_t = h_p + h_s$$

$$h_p = \sqrt{\frac{P}{24.5H}}$$

$$h_s = KP\frac{dh}{dP}$$

$$= \frac{KP\sqrt{\pi}}{2h_p \beta E^* \sqrt{24.5}} \qquad (A1.6.1a)$$

$$= \frac{K\sqrt{P}\sqrt{\pi}}{2h_p \beta E^*} \frac{\sqrt{PH}}{\sqrt{24.5H}}$$

$$= \frac{K\sqrt{PH\pi}}{2\beta E^*}$$

$$h_t = \sqrt{P}\left[\frac{1}{\sqrt{24.5H}} + \frac{K\sqrt{H\pi}}{2\beta E^*}\right]$$

For a purely elastic response, the relationship between depth and load (for a Berkovich indenter) is:

$$P = \frac{2}{\pi}h_t^2 E^* \tan\alpha$$

$$h_t = \sqrt{P}\left[\sqrt{\frac{\pi}{2E^* 2.174}}\right] \qquad (A1.6.1b)$$

The significance of these results is that the measured total indentation depth h_t is proportional to the square root of the load for both elastic and plastic deformation, the constant of proportionality of course being different.

If E^* and H are known, then the complete load versus depth history, for loading, of the plastic deformation of an ideal specimen can be generated using Eq. A1.6a. *For the purposes of analysis*, for the elastic unloading, the relationship between depth and force can be assumed linear according to dP/dh at the maximum load. The actual material response during unload may *initially* follow the power law proposed by Oliver and Pharr.

A1.6.2 Spherical indenter

Simulation of the loading and partial unloading curves for a spherical indenter is possible if it is assumed that the radius of the circle of contact at full load and full unload are identical. The consequences of this assumption are described in Chapter 10. The method described here is that formulated by Field and Swain[5]. Upon the application of load during an indentation test there is generally an initial elastic response followed by elastic plastic deformation. For the purposes of simulation, the transition between the two responses is assumed to occur at a mean contact pressure equal to the hardness H. That is, we are assuming that there is an abrupt transition from elastic deformation to a fully developed plastic zone and no intermediate region (see Chapter 9). The data to be calculated is the load and depth for both loading and each partial unloading sequence. At low loads, the specimen response is elastic and the relationship between depth and load is the familiar Hertzian expression:

$$h_t^3 = \left[\frac{3}{4E^*}\right]^2 \frac{P^2}{R_i} \qquad (A1.6.2a)$$

For a rigid indenter of radius R_i, the depth of penetration beneath the specimen free surface is also given by:

$$h_t = \frac{a^2}{R_i} \qquad (A1.6.2b)$$

and thus:

$$h_t = \frac{3}{4E^*}\frac{P}{a} \qquad (A1.6.2c)$$

When a critical load P_c is reached, full plasticity is assumed and $p_m = H$. Thus, with $p_m = P/\pi a^2$, it can be shown from Eq. A1.6.2b and A1.6.2a that the critical load is given by:

$$P_c = \left(\frac{3}{4E^*}\right)^2 (\pi H)^3 R_i^2 \qquad (A1.6.2d)$$

At the load P_c, there is a corresponding radius of circle of contact a_c which is found from the hardness H:

$$a_c = \sqrt{\frac{P_c}{\pi H}} \qquad (A1.6.2e)$$

Beyond the critical load, no further increase in the mean contact pressure occurs with increasing depth (H is a constant). We wish to calculate the depth of penetration beneath the specimen free surface as a function of indenter load given values of E^* and H for the material. The first step is to calculate the radius of the circle of contact from the hardness and the load. This can be expressed in terms of the radius of the circle of contact at the critical load:

$$a = a_c \left(\frac{P}{P_c}\right)^{\frac{1}{2+x}} \qquad (A1.6.2f)$$

In Eq. A1.6.2f, x is a strain hardening index which accounts for any increase in the value of H with increasing load. For a constant value of H, x=0.

On unloading, we can assume that the response is elastic from the depth at full load, h_t, to the final residual depth h_r. Thus, inserting Eq. A1.6.2e into A1.6.2c, the elastic displacement, h_e, may be expressed by:

$$h_e = \frac{3}{4E^*} \frac{P}{a_c} \qquad (A1.6.2g)$$

If we now assume that the radius of the circle of contact at full load P lies halfway between h_t and h_r, then:

$$h_t = \frac{2h_p + h_e}{2} = 2h_p - h_r \qquad (A1.6.2h)$$

and thus:

$$h_t = \frac{1}{2}\left(2h_p + \frac{3}{4E^*} \frac{P}{a_c}\right) \qquad (A1.6.2i)$$

Equation A1.6.2g expresses h_t in terms of h_p. Now, from geometry, the distance h_p is also given by:

$$h_p = R_i - \sqrt{R_i^2 - a^2} \qquad (A1.6.2j)$$

where R_i is the radius of the indenter (assumed to be rigid) and a is given by Eq. A1.6.2f since we have assumed that the radius of the circle of contact is identical to that at full load. Substituting Eq. A1.6.2f into A1.6.2j, and then substituting Eq. A1.6.2j into Eq. A1.6.2i, we have the depth h_t as a function of E^*, R_i, H and the load P.

Appendix 1. Submicron Indentation Test Analysis

Equation A1.6.2a provides the loading curve up to a critical load P_c after which full plasticity is assumed. Equation A1.6.2i, with the appropriate substitutions, gives us an equation for the loading curve beyond the critical load P_c. The unloading curve is assumed to describe a fully elastic response from the total depth h_t to a residual depth h_r. However, this elastic unloading involves the indenter in contact with not a flat surface, but a preformed spherical impression (ie. the residual impression of radius R_r which remains upon the complete removal of load). Thus, we are able to use Eq. A1.6.2a for the unloading if R_i is replaced with the relative radius of curvature R given by:

$$\frac{1}{R} = \frac{1}{R_r} + \frac{1}{R_i} \qquad (A1.6.2k)$$

Now, R_r is unknown, but the radius of the circle of contact a at full load and the depth h_r is known (h_r can be calculated from Eq. A1.6.2h). However, we cannot simply use Eq. A1.6.2j to determine R_r from h_r and a since there we have not accounted for the assumptions implicit in the Hertzian equations used to simulate the loading part of the cycle. To be consistent, we need to effectively apply the same assumptions in reverse. Thus, we require then a value of R_r that would result in the known radius of circle of contact a at full load P when an indenter of radius R_i is loaded into a preformed impression through a distance h_e. Starting from Eq. A1.6.2a, we have:

$$h_e^3 = \left[\frac{3}{4E^*}\right]^2 \frac{P^2}{R} \qquad (A1.6.2l)$$

and:

$$h_e = \frac{a^2}{R} \qquad (A1.6.2m)$$

where a is the radius of the circle of contact at full load. Substituting Eq. A1.6.2m into A1.6.2l:

$$R = \frac{a^3}{P\frac{3}{4E}} \qquad (A1.6.2n)$$

and thus:

$$\frac{1}{R_r} = \frac{1}{R_i} - \frac{P\frac{3}{4E}}{a^3} \qquad (A1.6.2o)$$

from which we obtain:

$$R_r = \frac{R_i a^3}{a^3 - R_i P \dfrac{3}{4E}} \qquad (A1.6.2p)$$

Equation A1.6.2p gives the radius of the residual impression R_r required to produce a radius of circle of contact a when loaded elastically by a rigid indenter R_i through a depth h_e. The unloading curve is thus given by Eq. A1.6.2l with R the relative curvatures given by Eq. A1.6.2k. The absolute value of the unloading penetration (measured from the specimen free surface) is thus h_e (from Eq. A1.6.2l) added to the depth of the residual impression h_r.

References

1. M.F. Doerner and W.D. Nix, "A method for interpreting the data from depth-sensing indentation instruments," J. Mater. Res. 1 4, 1986, pp. 601–609.
2. W.C. Oliver and G.M. Pharr, "An improved technique for determining hardness and elastic modulus using load and displacement sensing indentation experiments," J. Mater. Res., 7 4, 1992, pp. 1564–1583.
3. G.M. Pharr, W.C. Oliver and F.R. Brotzen, "On the generality of the relationship among contact stiffness, contact area, and the elastic modulus during indentation," J. Mater. Res. 7 3, 1992, pp. 613–617.
4. R.B. King, "Elastic analysis of some punch problems for a layered medium," Int. J. Solids Struct. 23 12, 1987, pp. 1657–1664.
5. J.S. Field and M.V. Swain, "A simple predictive model for spherical indentation," J. Mater. Res. 8 2, 1993, pp. 297–306.

Appendix 2
The Finite-Element Method

A2.1 Introduction

The finite-element method is a numerical approximation technique used for the analysis of mechanical systems. Its main application is to analyze the response of materials to applied loads, but it also may be used to analyze heat transfer, fluid flow, vibration, electric potential, and other physical phenomena. A brief overview of the finite-element method is given in this chapter. Our purpose is to arrive at a very basic understanding of the theory behind the method and to become familiar with the way in which the method can be used to analyze contact problems.

A2.2 Finite-Element Analysis

How does the finite-element method work? We shall answer this question in relation to the analysis of applied loads on an elastic body. Generally, load may be applied to an elastic body which then deforms. To analyze the deflections and hence determine the induced stresses, the following steps are appropriate:

i. A set of differential equations is developed that apply to all points within the body.
ii. The set of equations is solved with appropriate boundary conditions applied.

The differential equations for the deflection of an elastic body under load are the equations of stress equilibrium and strain compatibility given in Chapter 1, Section 1.2.11 (Eqs. 1.2.11a and 1.2.11b). The solution of these equations often requires an approximation technique (e.g., truncated series, numerical integration) since geometry and boundary conditions are usually complex. The finite-element method is only one numerical approximation technique that can be used to solve these differential equations.

The method works by subdividing the geometry of the solid into "elements," which are connected together at "nodes" as shown in Fig. A2.2.1. It is assumed that the displacement of nodes for each element bears a "simple" relationship to other nodes (e.g., linear, quadratic).

A set of matrix equations is developed for each individual element. These matrix equations are assembled into a system of equations which then represent the complete structure. The system of equations is then solved numerically. In Eq. A2.2.1, ε is the strain associated with displacements u and in matrix notation, the strain is defined as:

$$[\varepsilon] = \begin{bmatrix} \dfrac{\delta}{\delta x} & 0 \\ 0 & \dfrac{\delta}{\delta y} \\ \dfrac{\delta}{\delta y} & \dfrac{\delta}{\delta x} \end{bmatrix} [u] \qquad \text{(A2.2.1)}$$

$$= [B][u]$$

where the matrix [ε] gives the strain in the x and y directions (in this "two-dimensional" example) and [u] is the corresponding matrix of displacements. Stress and strain are related of course by Hooke's law, which in matrix notation can be written:

$$[\sigma] = \dfrac{E}{(1+v)(1-2v)} \begin{bmatrix} 1-v & & \\ v & 1-v & \\ & & \dfrac{1-2v}{2} \end{bmatrix} [\varepsilon] \qquad \text{(A2.2.2)}$$

$$= [E][\varepsilon]$$

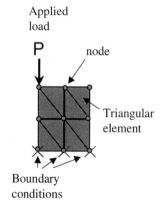

Fig. A2.2.1 Division of the geometry of a solid body into elements connected together at nodes.

Appendix 2. The Finite-element Method

If the entire structure is in equilibrium, then so is each element. Thus, for equilibrium:

$$\delta W = \delta U \qquad (A2.2.3)$$

where δW is the (virtual) work done by external loads and δU is the work done by internal forces. Thus, if δu is the displacement of a node i acted on by a force F_i, then at equilibrium:

$$[\delta u_i]^T [F_i] = \int_V [\delta \varepsilon]^T [\sigma] dV \qquad (A2.2.4)$$

where i = 1,2,3... up to the total number of elements in the model structure. The notation $[\ ]^T$ denotes the transpose of a matrix. Thus:

$$[\delta u_i]^T [F_i] = \int_V [\delta u_i]^T [B]^T [E][B][u_i] dV$$

$$[F_i] = \int_V [B]^T [E][B][u_i] dV \qquad (A2.2.5)$$

$$[F_i] = [K][u_i]$$

The matrix [K] is called the "element stiffness matrix."

$$[F_i] = [K][u_i]$$

$$[P_i] = [K][u_i] \qquad (A2.2.6)$$

The matrix [**K**] is the "structure stiffness matrix" and is formed by successively adding stiffness terms from each element into appropriate locations in the structure matrix. The matrices [**P**] are the applied loads. The procedure is thus:

i. Assemble stiffness matrix
ii. Apply constraints
iii. Solve matrix equation for the unknown displacements u_i

$$[u_i] = [K]^{-1}[P_i] \qquad (A2.2.7)$$

It can be appreciated from Eq. A2.2.7 that the solution for a finite-element analysis involves the inversion of a matrix, which can be quite large. It is no surprise that the popularity of the finite-element method has coincided with the development of readily available computers for this purpose.

There are generally three sources of error associated with the method. Discretization error occurs if the geometry of the actual structure is not adequately represented by the discrete assemblage of elements. Formulation error occurs when the element types are not suitable for the loading and geometry of the problem, and computational error occurs due to numerical round-off during matrix inversion. These errors can be minimized by a judicious use of the method which is a product of experience of the analyst. With the finite-element method it is very easy to obtain a result but not so easy to obtain the *correct* result.

A2.3 Finite-Element Modeling

The *elastic* stress fields generated by an indenter, whether it be a sphere, cylinder, or cone, although complex, are well defined. When the response of the specimen material is *elastic-plastic,* however, theoretical treatments (such as those mentioned in Chapter 9) are limited because of the simplifying assumptions required to make such analyses tractable. Although application of the finite-element method to elastic-plastic indentations with a spherical indenter is not new[1-9], it does require some specialist knowledge. The intention here is to present some background information to assist the analyst who wishes to undertake this complex modeling procedure.

A2.3.1 Contact between the indenter and the specimen

Contacts may generally be classified as either conforming or nonconforming[10]. Loading with a cylindrical flat punch indenter is an example of a conforming contact since the contact area is a constant and independent of the load. Loading with a spherical or conical indenter is a nonconforming contact since the contact area is dependent on the load. However, this does not mean that contact involving such an indenter is a nonlinear event. For frictionless contact, the contact area is completely specified in terms of linear elasticity as embodied in the equations presented in Chapter 6. Thus, as far as finite-element modeling is concerned, we should obtain the same solution whether the load is applied in one step or as a series of increments. The foregoing discussion assumes that all displacements are small. If this is not the case, then the elastic equations do not apply, and for finite-element modeling, nonlinear *geometric* considerations must be included in the analysis.

The solution of contact problems by finite-element analysis is often conveniently undertaken with the use of specialized gap elements. Although for frictionless contact, a linear solution is obtained if the full indenter load is applied at once, modeling of the expanding area of contact, such as that required for spherical and conical indenters, does require an iterative procedure. In the contact problem considered in this chapter, potential contacting surfaces are separated by specialized gap elements. The gap elements serve to prevent the "indenter" from overlapping the "specimen."

An iterative procedure continually checks the status of each gap element, deleting and reinstating the element as required, until force equilibrium is reached within a specified tolerance level. Ideally, the gap elements should be assigned an infinite stiffness. This would ensure a nonintrusive contact between the indenter and the specimen. However, this is not possible in practice due to the finite numeric restrictions imposed by computer hardware. During contact, the indenter intrudes the specimen a small amount, given by the penalty factor. The penalty factor is the ratio of the stiffness of the gap elements to the stiffness of the specimen material. It is desirable at least to have the stiffness of the gap beams considerably larger than that of the specimen material. If the penalty factor is too low, then there is insufficient stiffness to enforce the contact condition. A penalty factor of ≈10000 is usually sufficient to simulate contact in the present case.

A2.3.2 Elastic-plastic response

In the previous section, it was argued that indentation with a spherical indenter in frictionless contact with a flat specimen surface is essentially a linear contact problem, where full load may be applied at once and the expanding area of contact determined using an iterative procedure. Linear solutions of this type may be applied to a nonlinear material response by applying the load in increments, where the element response is assumed to be elastic within each increment. For material nonlinearity, the local elastic modulus of each element is modified at each iteration for each load increment to satisfy a specified constitutive relationship. This is commonly referred to as the "secant" method of nonlinear iteration. For example, the shear-driven nature of the subsurface contact damage observed in metals and some structural ceramics would suggest that the Tresca or von Mises shear stress yield criterion may be specified in the finite-element procedure.

The elastic-plastic properties of the specimen material in the finite-element model are specified by a uniaxial stress-strain relationship. For an elastic, perfectly plastic material, the stress-strain relationship is that shown by the heavy line in Fig. A2.3.1. Elastic-plastic behavior is accommodated by first applying load for the first increment with $E = E_o$ for all elements in the model. E_o is the gradient of the stress-strain curve, specified in the element property set, at zero strain. Within each load increment, starting with the first increment, iterations are performed until the specified tolerance level is reached. Within each iteration, the modulus of elasticity for each element is adjusted to satisfy the chosen yield criterion.

Figure A2.3.1 shows the procedure for a particular load increment. E_{now} is the initial stiffness for a particular element, which for the first iteration of the first load increment is E_o. In Fig. A2.3.1, this is shown as the stiffness prior to the application of the first load increment but may equally well represent the stiffness of a particular element after the analysis of a preceding load increment and series of iterations. For each element, the Tresca strain ε_{Tr} is calculated using E_{now}:

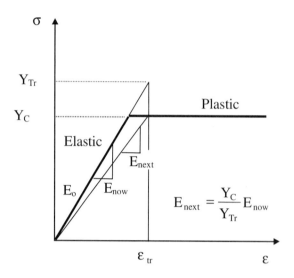

Fig. A2.3.1 Elastic-plastic stress-strain curve showing iterative procedure used to satisfy the yield criterion. The elastic modulus of each element is adjusted to satisfy the specified uniaxial stress-strain relationship, here shown as elastic-perfectly plastic (after reference 6).

$$\varepsilon_{Tr} = \frac{Y_{Tr}}{E_{now}} \tag{A2.3.2a}$$

where

$$Y_{Tr} = \max \left\| \sigma_1 - \sigma_2 \right|, \left| \sigma_2 - \sigma_3 \right|, \left| \sigma_3 - \sigma_1 \right\| \tag{A2.3.2b}$$

Y_{Tr} is compared with that obtained from the stress-strain curve Y_C at ε_{Tr}. If Y_{Tr} is greater than Y_C, then that element is carrying more stress than is permitted by the specified stress-strain curve and failure criterion. The modulus E_{now} for that element is then factored down an amount given by:

$$E_{next} = E_{now} \frac{Y_C}{Y_{Tr}} \tag{A2.3.2c}$$

and a new solution calculated for that load increment with $E = E_{next}$. Iterations are performed until the values for Y_{Tr} agree, within a specified tolerance level, with those determined from the curve at a given ε_{Tr}.

The approach to plasticity described above is adequate for the purposes of modeling the application of load but is not suitable when one wishes to analyze the unloading. For unloading, more sophisticated techniques such as the incremental strain or incremental stress methods[11] are required.

A2.3.3 Finite-element model

A schematic of a portion of the finite-element model used to obtain the results presented in Chapters 8 and 9 (Figs. 8.3.1–8.3.3 and 9.3.6–9.3.7) is shown in Fig. A2.3.2. The complete finite-element mesh consists of 1736 nodes and 1538 axis-symmetric quadrilateral plate elements. The outer dimensions of the model are some 150 times the radius of the circle of contact at the highest load increment.

To enable direct comparison with experimental results, the indenter radius was set to 3.18, mm which resulted in a node spacing directly beneath the indenter of ≈ 8 µm. The stiffness of the gap elements was set larger than that of the elements representing the specimen by a factor of $\approx 1 \times 10^6$. During the solution, a maximum of 25 iterations per load increment were specified, and this was sufficient to obtain convergence with a force tolerance of 1% of the normalized residual force and 0.5% of the maximum normalized displacement. Load was applied in a series of steps up to the maximum specified load. A total of 20 load increments were specified. An updated Lagrangian method was employed within the finite-element code to account for the geometric nonlinearity associated with large strains which may arise in the material in which yield has occurred.

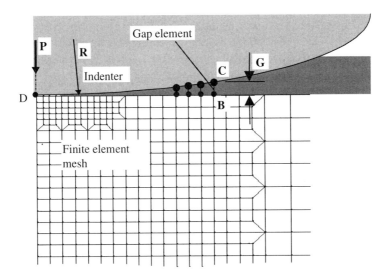

Fig. A2.3.2 Detail of finite-element mesh near prospective contact area. Four gap-beam elements are shown. Constraints and freedoms for nodes at C and A ensure that nodes at B and C move together when gap G closes during application of load (after reference 6).

Material properties, such as elastic modulus and yield stress, for the specimen material were determined by experiment using the indentation stress-strain response of the material. The elastic modulus E is found from the linear portion of the indentation stress-strain curve, and the yield stress Y may be estimated from the value of mean contact pressure corresponding to the point of first deviation from linearity on such a curve.

A2.3.4 Finite-element modeling results

Considerable care needs to be taken in the interpretation of the results of a finite-element analysis. All too often, the problem is that there is an excess of data from which the features of interest are to be extracted. In essence, the analysis calculates the displacements of nodes as a result of applied loads and boundary conditions. Most finite-element analysis packages present structural output in terms of stresses and strains, which are more useful in engineering analysis.

One of the issues that requires caution is the orientation of the elements comprising the model. Elements in a finite-element model generally have a local coordinate system, which may not coincide with the overall, or global, coordinate system of the geometry of the structure. Local coordinate systems are often used to specify anisotropic material properties. The nature of the geometry of the structure may require that the stresses obtained from a finite-element solution be expressed relative to planes aligned with each element's local coordinate axes. For example, an analysis of the hoop stress in a pipe of circular cross-section under internal pressure would require stresses to be expressed relative to a local coordinate axis for each element. Global stresses in this example would be meaningless. Contact problems are generally axis-symmetric and hence global stresses in the radial, hoop, and axial (or "normal") directions are of interest no matter what the orientation of the elements within the model.

Another potential problem is the averaging technique employed by the particular finite-element analysis code. Consider the results for a 4-node plate element. Typically, stresses, strains, and displacements are reported for each node. However, internally, the program may actually calculate the solutions at the centroids of each area shown and provide a weighted average nodal output. If the nodes of interest lie on a boundary (say between elements with two different material properties), then this averaging may reduce the actual stresses at the boundary where the results are required. In some cases, it may be possible to specify that averaging is performed only within elements of the same material property.

Often, a simple display of stresses is sufficient for a particular analysis. Stress concentrations may be readily identified and future modeling performed on a revised geometry or loading as required. Stress fields for an elastic-plastic indentation of a flat specimen with a spherical indenter have been given in Chapter 8 (see Figs. 8.3.1 to 8.3.3). In contact problems, the stresses along the axis of symmetry and the specimen surface are of considerable importance and may be readily extracted from the finite-element solution file.

A more sophisticated output may be obtained with only a little more effort. In this book, we have discussed the importance of indentation stress-strain and compliance curves. An indentation stress-strain curve is obtained by plotting values of mean contact pressure for a spherical indenter against the ratio a/R, where a is the radius of the circle of contact and R is the radius of the spherical indenter (see Fig. 9.3.8). There is scope for error in this procedure when dealing with finite-element results. Contacts can only occur between nodes, and the nodes are a discrete distance apart. Thus, the most accurate estimate of the radius of circle of contact a would be the average radial coordinates of the node that has contacted and the next one farther out from the axis, which has not yet contacted. Since the mean contact pressure involves a $1/a^2$ term, and the indentation strain depends directly on a, any error in the estimation of radius of circle of contact is cubed. Compliance curves (Fig. 10.4.5) are obtained from the displacement of the indenter as a function of applied load. Information of this type may require many finite-element analyses to be performed (one for each loading increment) so that a complete picture of the response of the material may be calculated.

References

1. C. Hardy, C.N. Baronet, and G.V. Tordion, "The elastic-plastic indentation of a half-space by a rigid sphere," Int. J. Numer. Methods Eng. 3, 1971, pp. 451–462.
2. P.S. Follansbee and G.B. Sinclair, "Quasi-static normal indentation of an elasto-plastic half-space by a rigid sphere-I," Int. J. Solids Struct. 20, 1981, pp. 81–91.
3. R. Hill, B. Storakers and A.B. Zdunek, "A Theoretical study of the Brinell hardness test," Proc. R. Soc. London, Ser. A423, 1989, pp. 301–330.
4. K. Komvopoulos, "Finite element analysis of a layered elastic solid in normal contact with a rigid surface," Trans. ASME 111, 1988, pp. 477–485.
5. K. Komvopoulos, "Elastic-plastic finite element analysis of indented layered media," Trans. ASME 111, 1989, pp. 430–439.
6. G. Carè and A.C. Fischer-Cripps, "Elastic-plastic indentation stress fields using the finite element method," J. Mater. Sci. 32, 1997, pp. 5653–5659.
7. J. Tseng and M.D. Olson, "The mixed finite element method applied to two-dimensional elastic contact problems," Int. J. Numer. Methods Eng. 17, 1981, pp. 991–1014.
8. N. Okamoto and M. Nakazawa, "Finite element incremental contact analysis with various frictional conditions," Int. J. Numer. Methods Eng. 14, 1979, pp. 337–357.
9. T.D. Sachdeva and C.V. Ramakrishnan, "A finite element solution for the two-dimensional elastic contact problems with friction," Int. J. Numer. Methods Eng. 17, 1981, pp. 1257–1271.
10. K.W. Man, *Contact Mechanics using Boundary Elements*, Computational Mechanics Publications, Southamton, UK, 1994.
11. O.C. Zienkiewicz and Y.K. Cheung, *The Finite Element Method in Structural and Continuum Mechanics*, McGraw-Hill, London, 1967.

Index

abrasion, 49
Auerbach, 118, 119, 120, 121, 123, 124, 125, 126, 127, 128, 130, 131, 132, 133, 134, 136, 137, 157, 174
Auerbach range, 123, 124, 125, 126, 127, 128, 130, 132, 134
axial symmetry, 22

barreling, 157
Berkovich, 139, 152, 186
biaxial stress, 71, 72
blister field, 142, 143
bonded interface, 202, 203
Boussinesq, 78, 80, 81, 83, 93, 101, 116, 142, 145
Brinell, 152, 153, 154, 174, 179

Charles and Hillig, 50, 51, 52, 53, 54, 56, 60
chemical bonds, 1, 5
clip gauge, 207, 208
cohesive strength, 2, 181
compliance, 173, 185, 192, 208
compression, 5, 6, 10, 15, 18, 25, 29, 84, 104, 158, 161, 168, 174, 198, 204, 208
compressive zone, 204
concentrated force, 79
cone cracks, 92, 103, 116, 117, 118, 210
confining pressure, 157, 158, 204
Coulomb forces, 1
Coulomb repulsion, 1
crack growth, 33, 34, 35, 43, 46, 49, 50, 54, 57, 58, 61, 126, 133, 136
crack path, 135
crack resistance, 41, 75

crack tip, 33, 34, 35, 36, 37, 39, 40, 41, 42, 43, 44, 45, 47, 50, 51, 52, 53, 54, 92, 122, 135, 148
crack tip plastic zone, 40, 53
crack velocity, 50, 51, 53, 54
cryogenic temperatures, 50
cumulative probability distribution, 62, 63
cylindrical flat punch, 77, 79, 93, 96, 97, 109, 124, 131, 133, 181, 185
cylindrical roller, 92

dead-weight, 36
delayed failure, 74
densification, 142, 168
deviatoric components, 25, 26, 28, 29
deviatoric stress, 8, 25, 29
dilatation, 9
dissipative mechanisms, 34, 36, 41, 172
distance of mutual approach, 88, 106, 107, 108, 109, 110
distribution of pressure, 77, 78, 79, 88, 93, 107, 165

elastic constraint, 153, 161, 165, 172, 181, 182, 185
elastic modulus, 1, 3, 18, 105, 108, 111, 148, 196, 199, 230
elastic recovery, 171, 182, 183, 185
elastic stress fields, 77, 139
energy balance criterion, 34, 35, 38, 41, 43, 46, 49, 61, 119, 120, 121
environment, 49, 50, 51, 54, 144
equilibrium, 1, 2, 3, 4, 7, 16, 23, 24, 35, 36, 39, 40, 42, 77, 136, 137
estimator, 70, 71

242   Index

expanding cavity, 140, 142, 145, 161, 165, 168, 169, 173, 181

finite element, 32, 46, 47, 86, 145, 148, 149, 150, 151, 168, 171, 196, 197
fixed-grips, 36
flaw statistics, 120, 134, 135
fluid flow, 27
fracture surface energy, 33, 36, 119, 120, 124, 125, 130, 131, 132
fracture toughness, 146
friction, 112
frictionless contact, 77, 105, 127, 155, 166

geometrical similarity, 164, 167, 177, 178
Griffith, 32, 33, 34, 37, 41, 43, 46, 48, 49, 54, 59, 61, 68, 119, 120, 121, 122, 124, 125, 127, 128, 129, 135, 136
Griffith criterion, 61

Hertz, 77, 78, 87, 101, 103, 104, 107, 108, 116, 117, 120, 121, 125, 127, 132, 136, 137, 153, 157, 159, 171, 174, 197, 203
Hertzian cone cracks, 90, 103, 117, 118, 122, 137
Hooke's law, 2, 80
hydrostatic, 25, 28, 29, 151, 158, 162, 163

impact, 110
indentation fracture, 144
indentation strain, 104, 159, 160, 161, 169, 170, 173, 179, 203, 204, 205, 206
indentation stress, 13, 23, 30, 77, 78, 83, 90, 92, 95, 104, 105, 110, 121, 122, 127, 129, 139, 140, 148, 150, 153, 158, 159, 165, 169, 171, 180, 202, 203, 204, 205, 206
inert strength, 210
Inglis, 31, 32, 33, 48, 49, 50, 51, 54, 59
interfacial friction, 112
internal pressure, 140, 161
Irwin, 37, 41, 48, 50, 53, 54, 59, 61, 119, 136

K_{IC}, 41, 42, 44, 50, 54, 55, 61, 74, 119, 122, 127
Knoop, 139, 155, 156, 174, 214, 215

lateral cracks, 144, 145
Laugier, 147, 148, 152
linear elasticity, 15, 40, 71, 159, 204
load-point displacement, 106, 107, 108, 110, 192, 207, 208

mean contact pressure, 87, 91, 94, 95, 96, 97, 100, 104, 108, 109, 110, 122, 140, 141, 143, 148, 149, 154, 155, 158, 159, 161, 164, 165, 171, 178, 179, 180, 181, 203, 204, 212
median cracks, 144
Meyer, 154, 157, 174
minimum critical load, 125, 127, 128, 130, 132, 134

Navier–Stokes, 27
Nomarski, 201, 202

Obreimoff's experiment, 36

Palmqvist, 144, 146, 151, 152
permanent set, 157
plane strain, 16, 17, 20, 21, 29, 30, 33, 42, 43, 44, 50, 119
plane strain fracture toughness, 42, 44, 50
plane stress, 16, 17, 19, 29, 42, 43, 44
plastic zone, 40, 41, 42, 47, 53, 54, 140, 141, 143, 144, 150, 160, 161, 162, 163, 165, 166, 169, 170, 172, 173, 181, 183
plasticity, 27, 28, 54, 139, 148, 153, 161, 165, 173, 182, 183, 185, 197, 198, 208
point contact, 80
point-load, 79
Poisson's ratio, 14
potential energy, 2, 4, 5, 11, 33, 36, 38, 42, 111
principal planes, 18, 19, 20, 21, 22, 23, 28
principal stresses, 18, 19, 20, 21, 22, 23, 26, 29, 30, 71, 72, 90, 93, 98, 148, 158, 170

prior stress field, 46, 121, 126, 135
probability density function, 63, 64, 66
probability of failure, 64, 65, 66, 67, 68, 69, 70, 71, 72, 75, 126, 127, 128, 129, 130, 131, 132, 133, 134
proof stress, 56, 57, 58, 74

radial cracks, 144
relative radii of curvature, 107
reloading, 183
rigid conical indenter, 97
rigid-plastic, 140, 153, 160, 166, 181
Rockwell, 156
Roesler, 136, 137, 151

SENB, 44
sharp tip crack growth model, 56, 57
shear modulus, 15, 82
shear strains, 10
shear stresses, 5, 7, 8, 9, 12, 15, 18, 20, 22, 26, 28, 30, 92, 95, 197
shearing angle, 11, 12, 13
Shore scleroscope, 156
slip, 112
slip-line theory, 168, 181
soda-lime glass, 50, 119, 127, 131, 132, 133, 141, 144, 150
spherical indenter, 87, 105, 148
St. Venant's principle, 25, 140
static fatigue, 49
static fatigue limit, 50, 54, 58, 74, 75
strain, 10
strain compatibility, 23
strain energy, 4, 5, 29, 33, 34, 35, 36, 38, 40, 41, 42, 46, 61, 111, 119, 121, 122, 123, 125, 126, 127, 128, 135, 136, 172, 173, 185
strain energy release rate, 34, 36, 40, 42, 43, 46, 119, 121, 122, 123, 125, 127, 128

stress concentration factor, 31, 32, 33, 50, 51, 52, 54
stress corrosion theory, 50, 51, 54, 56
stress deviations, 8, 25
stress equilibrium, 23
stress intensity factor, 37, 39, 41, 42, 43, 44, 45, 46, 47, 50, 53, 54, 55, 58, 61, 74, 119, 121, 135, 147
stress trajectories, 26, 81, 91, 96, 100
subcritical crack growth, 50, 54, 56, 57, 58, 61, 74
surface energy, 4, 5, 33, 34, 35, 36, 41, 42, 49, 61, 131, 132, 133, 134
surface tractions, 44, 45, 46

tensile strength, 3, 31, 35, 53, 64, 66, 68, 119, 154, 211, 212
tension, 5, 6, 10, 15, 18, 25, 28, 29, 31, 43, 64, 84, 104, 140, 210
tensor, 8, 12, 13
time to failure, 54
Tresca, 28, 30, 159, 166, 167, 181, 198, 204

uniform pressure, 84, 105

Vickers, 139, 144, 145, 146, 147, 152, 155, 168, 178, 180, 181, 205, 207, 212, 213, 215
Vickers diamond pyramid, 139, 147, 168, 180, 181, 205, 207, 213
viscosity, 27
von Mises, 29, 30, 166, 167, 204

Weibull parameters, 66, 67, 70, 71, 72, 127, 130, 133
Weibull statistics, 61, 64, 126, 150

yield stress, 26, 28, 29, 40, 158, 159, 165, 168, 181, 183, 204
Young's modulus, 3

Mechanical Engineering Series *(continued from page ii)*

D.P. Miannay, **Fracture Mechanics**

D.K. Miu, **Mechatronics: Electromechanics and Contromechanics**

D. Post, B. Han, and P. Ifju, **High Sensitivity Moiré: Experimental Analysis for Mechanics and Materials**

F.P. Rimrott, **Introductory Attitude Dynamics**

S.S. Sadhal, P.S. Ayyaswamy, and J.N. Chung, **Transport Phenomena with Drops and Bubbles**

A.A. Shabana, **Theory of Vibration: An Introduction, 2nd ed.**

A.A. Shabana, **Theory of Vibration: Discrete and Continuous Systems, 2nd ed.**